BARNES & NOBLE PRES

T0256417

OUR 3RD ANNUAL

Mini Maker Faire®

IN PARTNERSHIP WITH **MAKER MEDIA**, CREATORS OF **MAKE MAGAZINE** & **MAKER FAIRE**

NOV 11 + NOV 12

FOR ALL AGES

BARNES & NOBLE
Mini **Maker Faire**

MAKE WORKSPACE
Experience the latest in virtual reality, designing, robotics, coding, programming & more.

MEET THE MAKERS
Hear from the leaders and top brands about the latest trends in designing, building, creating & making.

MAKE & COLLABORATE
Immerse yourself in the process of ideating, creating & constructing with a vast array of materials ideally suited for making anything possible.

 TO FIND A PARTICIPATING STORE, VISIT **BN.COM/makerfaire**   **#BNMake2017**

Barnes & Noble Is a Proud Presenting Sponsor of World Maker Faire
New York Hall of Science in Corona, NY SEPT 23 + SEPT 24

CONTENTS

Make: **Volume 59** October/November 2017

HOME HACKS

ON THE COVER:
One of the diminutive dwellings by Jay Shafer's Four Lights Tiny House Company invites you to come in and get cozy.

Photo: Hep Svadja.

Paul Bartlett, Blank William, Lee Camara, Goli Mohammadi, Hep Svadja, Rob Nance

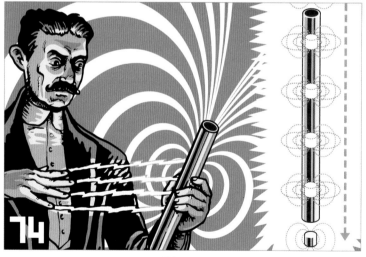

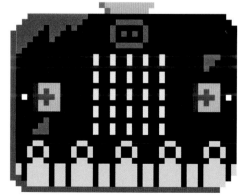

EXECUTIVE CHAIRMAN & CEO
Dale Dougherty
dale@makermedia.com

CFO & PUBLISHER
Todd Sotkiewicz
todd@makermedia.com

VICE PRESIDENT
Sherry Huss
sherry@makermedia.com

EDITORIAL

EXECUTIVE EDITOR
Mike Senese
mike@makermedia.com

SENIOR EDITOR
Caleb Kraft
caleb@makermedia.com

EDITOR
Laurie Barton

MANAGING EDITOR, DIGITAL
Sophia Smith

PRODUCTION MANAGER
Craig Couden

EDITORIAL INTERN
Jordan Ramée

CONTRIBUTING EDITORS
William Gurstelle
Charles Platt
Matt Stultz

CONTRIBUTING WRITERS
Alasdair Allan, Bonnie Burton, Kathy Ceceri, Jon Christian, Larry Cotton, DC Denison, Stuart Deutsch, Russell Graves, Mark Longley, Brian Lough, Lisa Martin, Goli Mohammadi, Aaron Newcomb, Jeff Olson, Max Ritter, Karl Sims, Julia Skott, Bruce Sterling, Mandy L. Stultz, Jason Suter, Gordon Williams, Alex Wulff, Chris Yohe, Bill Young

DESIGN, PHOTOGRAPHY & VIDEO

ART DIRECTOR
Juliann Brown

PHOTO EDITOR
Hep Svadja

SENIOR VIDEO PRODUCER
Tyler Winegarner

LAB INTERN
Luke Artzt

MAKEZINE.COM

DIRECTOR, PRODUCT & ENGINEERING
Jared Smith

TECHNICAL PROJECT MANAGER
Jazmine Livingston

WEB/PRODUCT DEVELOPMENT
David Beauchamp
Bill Olson
Sarah Struck
Alicia Williams

CONTRIBUTING ARTISTS
Matthew Billington, Monique Convertito, Rob Nance, Peter Strain

ONLINE CONTRIBUTORS
Alexis Gabriel Ainouz, Dennis Atwood, Brian Berletic, Gareth Branwyn, Chiara Cecchini, Jeremy Cook, Leo Deluca, Liam Grace-Flood, Jess Hobbs, Homemade Game Guru, Elijah Horland, Brian Jepson, Shawn Jolicoeur, Hee Jung, Cristoph Klemke, Becky LeBret, Bill Livolsi, Jane Lyons, Richard Ozer, Sydney Palmer, Angela Perrone, Sandra Rodriguez, Toshinao Ruike, Kerry Scharfglass, Maret Sotkiewicz, Ben Vagle, Glen Whitney, Dan Woods, Frank Zhao, Lee D. Zlotoff

PARTNERSHIPS & ADVERTISING
makermedia.com/contact-sales or partnerships@makezine.com

DIRECTOR OF PARTNERSHIPS & PROGRAMS
Katie D. Kunde

STRATEGIC PARTNERSHIPS
Cecily Benzon
Brigitte Mullin

DIRECTOR OF MEDIA OPERATIONS
Mara Lincoln

BOOKS

PUBLISHER
Roger Stewart

EDITOR
Patrick Di Justo

PUBLICIST
Gretchen Giles

MAKER SHARE

DIRECTOR, ONLINE OPS
Clair Whitmer

CONTENT & COMMUNITY MANAGER
Matthew A. Dalton

LEARNING EDITOR
Keith Hammond

MAKER FAIRE

EXECUTIVE PRODUCER
Louise Glasgow

PROGRAM DIRECTOR
Sabrina Merlo

MARKETING & PR
Bridgette Vanderlaan

COMMERCE

PRODUCTION AND LOGISTICS MANAGER
Rob Bullington

PUBLISHED BY

MAKER MEDIA, INC.
Dale Dougherty

Comments may be sent to:
editor@makezine.com

Visit us online:
makezine.com

Follow us:
🐦 @make @makerfaire @makershed
google.com/+make
makemagazine
makemagazine
makemagazine
twitch.tv/make
makemagazine

Manage your account online, including change of address:
makezine.com/account
866-289-8847 toll-free
in U.S. and Canada
818-487-2037,
5 a.m.–5 p.m., PST
cs@readerservices.makezine.com

Issue No. 59, October/November 2017. Make: (ISSN 1556-2336) is published bimonthly by Maker Media, Inc. in the months of January, March, May, July, September, and November. Maker Media is located at 1700 Montgomery Street, Suite 240, San Francisco, CA 94111. SUBSCRIPTIONS: Send all subscription requests to Make:, P.O. Box 17046, North Hollywood, CA 91615-9588 or subscribe online at makezine.com/offer or via phone at (866) 289-8847 (U.S. and Canada); all other countries call (818) 487-2037. Subscriptions are available for $34.99 for 1 year (6 issues) in the United States; in Canada: $39.99 USD; all other countries: $50.09 USD. Periodicals Postage Paid at San Francisco, CA, and at additional mailing offices. POSTMASTER: Send address changes to Make:, P.O. Box 17046, North Hollywood, CA 91615-9588. Canada Post Publications Mail Agreement Number 41129568. CANADA POSTMASTER: Send address changes to: Maker Media, PO Box 456, Niagara Falls, ON L2E 6V2

CONTRIBUTORS

Where is the oddest place you keep tools in your home?

Bonnie Burton
San Francisco, CA
(Clever Couture)
The oddest place I have all my tools and craft supplies is well organized in their boxes and jars. It's the only place in my house that I actually keep tidy. The rest of my home looks like a geek episode of *Hoarders*.

Goli Mohammadi
Forestville, CA
(Heirloom Tech)
In my basement office, I have a small closet that converts into a snowboard tech station, complete with waxing iron, edge-sharpening files, P-Tex, scrapers, and base brushes.

Karl Sims
Cambridge, MA
(Three-Pendulum Harmonograph)
In the kitchen there is often duct tape, Gorilla Glue, and an electric drill/screwdriver, even though my wife disapproves when the kitchen table is used as a workbench.

Jeff Olson
Denver, CO
(DIY Hover Plant)
Shop Vac in the entry way closet, it's just better. Toolbox on the living room end table, no good reason. Allen wrenches under the bed because Ikea.

Jason Suter
Cape Town, South Africa (Blooming Flower Night Light)
The liquor cabinet usually contains at least glues, screwdrivers and scalpels, along with kids' toys to be fixed and 3D prints to be trimmed whenever the little folk allow it.

Lots of Love for Limor

Twitter users celebrate Adafruit's Fried on the cover of *Make:* Vol. 57

Carol Willing @WillingCarol · Follow

So cool that @make is featuring one of the best engineers ever on the cover of #makev57. Create and educate!

Make: @make
Announcing the next issue of Make, with our 2017 Boards Guide and a profile of @adafruit CEO Limor Fried. Subscribe: readerservices.makezine.com/mk/default.asp...

5:46 PM · 3 May 2017

Jay @Snabbjim6 · Follow

Finally got a copy of @make's #makev57 with @adafruit Ladyada on the cover!! (Also my first ever Make magazine purchase)

annereil @annereil · Follow

@make when does #57 with @adafruit hit the newstands!??!?! -- This is a MUST HAVE!!!!!

5:32 PM · 10 May 2017

James Steven Dunham @dugout · Follow

#makev57 Lady Ada Rocks.. Great Cover... Good job Make:

5:03 AM · 4 May 2017

Doctor Voom @docvoom · Follow

@make your feature on Lady Ada is fantastic. We need roll models for young makers like my daughter. #MakeV57

5:48 PM · 3 May 2017

Bruce Bufford @BruceBufford · Follow

@make @adafruit #MakeV57 So awesome seeing women engineers on the cover! Insta-purchase! Gotta get it autographed! :)

5:48 PM · 3 May 2017

Darrell Little @Darrell_VA · Follow

LadyAda on the cover of #MakeV57 greatly improved my opinion of @make magazine. Time to renew my subscription! #awesomeness #DIY #makers

5:51 PM · 3 May 2017

Eyes Wide Open @Scott205 · Follow

Finished my current edition of #Makev57 with Lady Ada @adafruit cover. Among the best issues I've ever read. Serious collector's item.Thanks

7:11 AM · 16 May 2017

A.T. Makers @at_makers · Follow

Found it! #makev57 right up front at @BNBuzz in Clearwater thanks for featuring great women like @adafruit!

11:45 AM · 3 Jun 2017

Jeff @JeffOnTheShelf · Follow

Love seeing Ladyada on cover of @make! I can't wait to get a copy. I absolutely love @adafruit #makev57

6:50 PM · 3 May 2017

rosmi @rosmi · Follow

#makev57 love ya.

8:58 PM · 20 May 2017

Make: Amends

In *Make:* Vol. 58, we misspelled the name of Roaming Tardis co-creator Bob Berger (page 43). Sorry Bob!

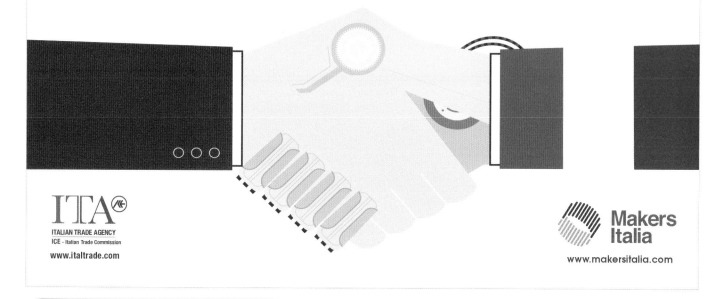

Future of Home

BY MIKE SENESE, executive editor of *Make:* magazine

Hep Svadja

Makers are an artful bunch, pursuing varied projects that go well beyond the benchtop. Their output often delves into the concept of home, using their dwellings as makerspaces, embracing and hacking connected devices, and even reimagining how residences are designed and inhabited. It's simply in makers' nature to tinker with and examine all facets of their worlds, including the places they live, and that's what we explore in this issue.

You can find the community's home personalization efforts in various venues — Maker Faire Rome has maybe the broadest variety I've seen of all. Alongside the Arduino rooms and 3D printer lounges, that

faire includes a large expo of fabricated furniture, open source kitchens, connected appliances, art, and more, showing the unique ways that people throughout the region are using their digs as outlets for expression, learning, and creativity. But really, it's global.

When it comes to projects, I have adopted the approach that the common area of my home is ideal for making. Here's a tip: Keep your family's workshop on the main level of your house, rather than in the garage, basement, or shed, as it keeps the family together when projects are underway. With our electronics and fabrication areas, my son and his cousins always have hands-on activities

nearby where my wife and I can join in. It's fun and fast to jump into an idea this way.

With OpenSprinkler and the like, makers have also helped pioneer the now commercialized "connected home." I love the control that these devices have given us over our homes. There's something incredibly satisfying about seeing the house lights illuminate at night as they detect you approaching. I've triggered my lawn sprinklers with an app from the other side of the planet. And my family practically has a new roommate with our voice assistant. But for the positives these gadgets bring, they introduce new setbacks too. We now have to deal with

unannounced firmware updates leaving us in the dark, unable to watch the season premiere of *Game of Thrones* because the link between the connected remote, TV, receiver, and lights had quietly been broken (true story!). Debugging my suddenly nonfunctional smart devices has, more than once, made me scream at their lack of brains.

It's still a young world for these concepts, one that promises to improve with time and adoption and maker ingenuity. There are a lot of ways that our community is pushing all of these concepts forward; find inspiration in these pages, and be sure to share your future-home undertakings with us. You're setting the bar. ◐

FROZEN WEB-HEAD

This frozen treat may look delicious to eat in the summer heat, but just like Spidey, not everything is as it seems! This was a non-edible prop created by Disney XD to use as a network summer 'melty' graphic to promote the new Marvel's Spider-Man animated series.

MARVEL
SPIDER-MAN

ALL NEW SATURDAYS

MARVEL on XD

2D SPIDER-MAN DESIGN + 3D MODELING + PRINTING
2D drawings of Spidey are digitally designed and rendered into 3D models. They are engineered to be sturdy to withstand casting in ice/popsicle molding.

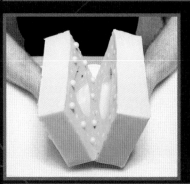

MOLDING + CASTING
The 3D prints are cleaned and sanded down before being cast in silicon to create the custom, one-of-a-kind Spidey molds for the popsicles.

POPSICLE CASTING
Sugar water is mixed with red dye to make the perfect Spider-Man red. This liquid is poured into the silicon molds, clamped shut and put in the freezer for twelve hours.

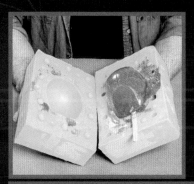

FINAL POPSICLES
Once frozen, the molds are carefully opened to reveal the 'tasty' treat. For a final frost, turn a can of air upside down, spray and voila!

MADE ON EARTH

Backyard builds from around the globe

Know a project that would be perfect for Made on Earth?
Let us know: *makezine.com/contribute*

THE FOREST AWAKENS

BLANKWILLIAM.COM

"The New Order project came from a doodle," says designer **William Kang**. He started with the rhino, and, since he likes to work in threes, added two other pachyderms: an elephant and a hippo. While Kang has done work in fashion, furniture, housewares, consumer electronics, and more, *Blank William* was a project he wanted to develop that he could have complete ownership of. Plus, he wanted a way to celebrate the release of *The Force Awakens*.

Kang uses a fluid process, working from sketches to multiple CAD packages to quick renders and back again, which allows him to progress toward a finished piece while still being able to backpedal and tweak details.

To make the physical sculpture, Kang worked with his fabricator friend **Vince Su**. They make silicone molds from 3D-printed parts, then mold wax prototypes that they dip into a ceramic slurry, reinforced with refractory sand. They then use the lost wax method and pour liquid metal into the mold. The helmets are stainless steel painted with an automotive finish, and they're supported with a metal stem fixed into a marble block. Each helmet takes a couple of months to make.

Kang says his interests jump from design and art to tech, fashion, pop culture, politics, science, and whatever else grabs his attention. He draws inspiration from everywhere, so he says the hardest part of his process is settling on an idea, as well as the communication and marketing.

—Sophia Smith

Kuan Yu-Hui

DIG THESE ROCKS

FACEBOOK.COM/PROFILE.
PHP?ID=100001801869745

José Manuel Castro López accomplishes the seemingly impossible: sculpting stone to look like soft putty. His artwork is stunning. There's no evidence that any of the stones have actually been worked on by tools.

It's almost a little unnerving. The stones look like they must have once been molten rock and López was simply lucky enough to find these beautiful creations. However, López points out that heat would destroy most of these rocks' internal structure.

"My stones don't endure any physical or chemical process," López says. "Behind every work there's a lot of drawing and modeling, that is to say, genuinely sculptural work." He continues, "The work must flow. Sometimes it stagnates, it collapses, it doesn't work out, it has to be fixed."

The Spanish sculptor cites his Galician heritage as the inspiration for his work. "In Galicia's culture, stone has always been mythical. My relation to stone isn't just physical, it's magical. I don't use the material as a mere support to sculpt around — we have a much more intimate relationship."

The result is an intriguing illusion that hovers between rigidity and malleability. Some of the stones have been sculpted to appear like the outer layer of the rock is being peeled away. Others look like López pinched two stones together, ripped a rock apart, or twisted a boulder to a point.

"I could say that [the stone and I] know each other, we understand each other, it obeys me, and reveals itself to me. In my works, there's no perseverance of the footprint of the sculptor. My stones are not sculpted, they manifest themselves."

López is making new sculptures all the time. Follow him on his Facebook page to see his entire gallery of work and get updates on his latest projects. —*Jordan Ramée*

José Manuel Castro López

ANXIETEA SARAHDUYER.COM

Sarah Duyer's series of teapots and cups seem as if they've stopped mid-crawl on their insectoid legs. This series is fittingly titled *Comfort/Discomfort*. By adding legs to objects that are traditionally tied to feelings of warmth and comfort, Duyer has not only changed the form, but the emotion evoked by the form is changed as well.

"Ceramic pieces are a part of our everyday lives and different forms can elicit different feelings and memories when they're used. That includes everything from your grandmother's china that sits precious and untouched in a cabinet until a special occasion, to your morning coffee mug that is a staple in your daily routine. In this series I wanted to explore the dichotomy of comfort and discomfort by altering those traditional, comforting forms and transforming them into something completely different," Duyer explains.

Adding the legs to the teapots also presented an interesting technical problem for the artist to solve. "I wanted legs that were strong enough to support the weight of the body but still visually convey its fragility," Duyer says. The challenge comes from the clay itself: even though clay is easy to mold, you are still limited by how fragile the piece will be once it is fired.

"The series took a lot of trial and error to figure how to attach the legs without them cracking or breaking from the weight of the teapot," says Duyer. The result is a series of teapots that are strangely imbued with character, and more than a little bit unsettling.

—Lisa Martin

Sarah Duyer

Megan Easterday - Flitetest

JUST PLANE COOL YOUTUBE.COM/AJW61185

Most people are fascinated by flight. Some bring that fascination into their workshop to build their own flying machines. For **Adam Woodworth**, that's not enough — his R/C creations take whimsical aircraft and spacecraft and bring them to life as actual flying toys. His latest creation is the classic Lego airplane, scaled up 10x from the original's wingspan of a few inches to almost six feet wide on his.

The San Francisco Bay Area-based hardware engineer has been interested in flight since before he can remember. "My dad was big into R/C planes in the '80s, so I basically grew up at the flying field," he says. Woodworth has been working in aerospace since college, while maintaining a prolific pace of model aircraft building — he estimates he's built several hundred aircraft over the past 25 years. "I was averaging a new build a month for a while," he explains.

After garnering considerable attention for his *Star Wars* and *Spaceballs* drone builds, Woodworth decided to focus on an area of interest for this specific build. "Lego is my #2 hobby and I've always wanted to do a sort of crossover project," he says. "When I found the giant 3D-printed Lego man, a plane for him to sit in was the obvious conclusion. It was really challenging to build it light enough. Since the shapes do not behave particularly efficiently, I had to keep things very light to work with the available lift."

The finished craft, weighing in under 4 pounds, flies surprisingly well despite its blocky wings with telltale studs. The craft is extremely accurate, even down to the Lego logos on those bumps, which Woodworth made using a 3D printed embossing stamp. He built the body from 3mm Depron and the wings from 1" EPS insulation sheet foam, both hand-cut and CNCed, over the course of 100 hours.

The build has been a hit. Woodworth says, "Everyone who sees the plane fly has a big smile, and that's what it's all about for me, sharing the joy of making and aviation." —*Mike Senese*

Expert cosplayers share their tips, tools, and ingenuity to inspire you to get started

Written by Bonnie Burton

Clever Couture

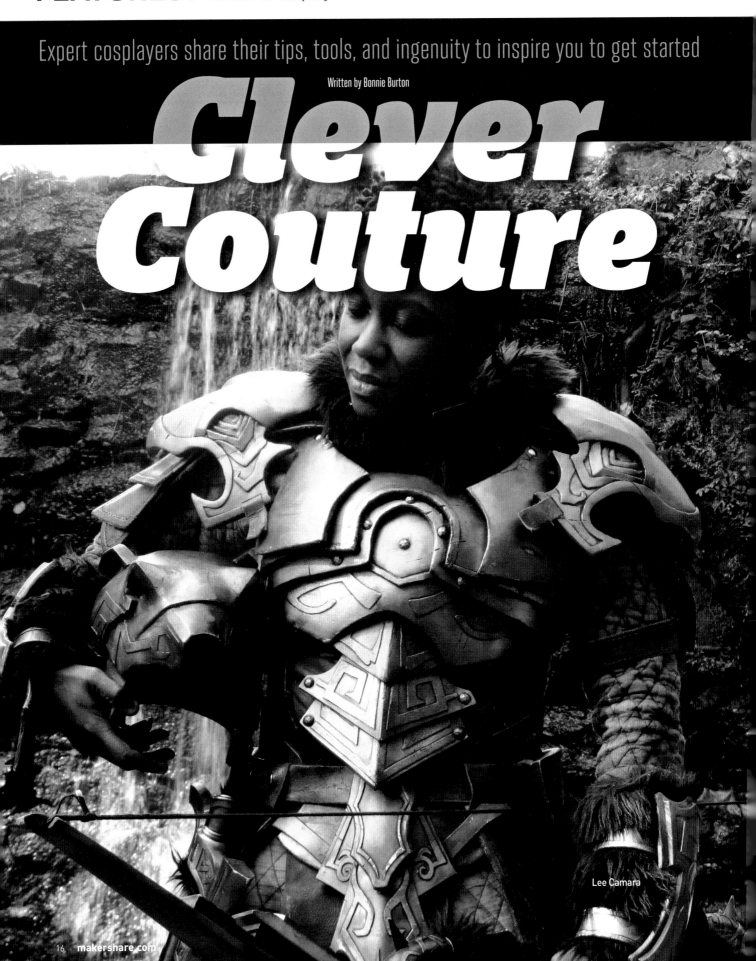

Lee Camara

Lee Camara

Lee Camara

Chloe Dykstra

Holly Conrad

Cosplay isn't just about posing for photos at conventions anymore. Cosplayers are impressing fans and makers alike by constructing their own original costumes that include elaborate armor, movie-ready prosthetics, special effects makeup, and intimidating weaponry.

Our favorite cosplayers — some of the most talented in the world — share their design and fabrication processes, favorite moments, and inspiring advice for newbies hoping to transform themselves. If you're bummed Halloween only comes once a year, this might be the hobby for you.

Lee Camara
fevstudios.com

My first "costume" was in 1996. I butchered some sweatpants to dress up as Haohmaru from *Samurai Spirits*. I later pulled out the costume when I went to my first convention in 1998. The convention bug bit me, and I attended them by participating in artist alleys. I began adding a prop or two as a display piece, and cosplayed if there was room to fit the costume in the suitcases. In 2004 I began doing commissions.

What's your process when making a costume?
I'll go through concept art and any other official images for reference. I'll sometimes make a hybrid of the design and sneak in personal touches.

I'll use Inkscape to create an orthographic drawing at full scale. If it needs to be

Henry Mei; Lee Camara, Holly Conrad, Paolo Cellammare, Greg de Stefano

sculpted, I'll just work from the reference images.

Additional sketches are done should the prop have special requirements, like the need to disassemble, safety,

Chloe Dykstra

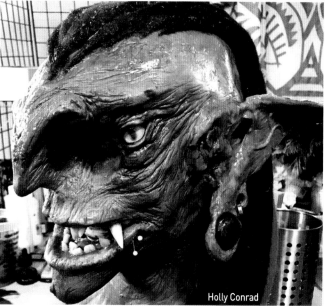
Holly Conrad

convention policies, weight, and other factors. Questions like "How quickly can I use the bathroom?" are also taken into consideration.

I have a laser cutter, and will cut some elements directly from the vector files. Most of the

work is done using hand and power tools. A Dremel, jigsaw, and my belt sander are the most frequently used, and the detail work is done with hand carving, sculpting, and filing tools.

For smaller accessories and sculpting in general, my preference is polymer clay. It can be baked in a toaster oven, carved, drilled, added upon, re-baked multiple times, and polished to a shine. Most items get molded and cast in resin.

What has been your favorite cosplay moment?
Meeting people who tuned in to my streams, or used tutorials to help them are my favorite. Seeing that it actually helped someone makes it all worth it.

Advice for aspiring cosplayers?
If you've never done a costume before, pick a smaller project that isn't too difficult. There are a ton of resources online. See if there are makerspaces or costuming groups that get together and build in your area.

Chloe Dykstra
chloedykstra.com

In 2010, my friends were making a web series called *There Will Be Brawl*, which was a dark re-imagining of Nintendo's lore. They asked me if I wanted to play Malon (from *Zelda: OoT*) as a prostitute. The only correct answer to that question is yes. I threw together a Malon cosplay from thrift store items and my closet, and that was the beginning of the end.

What's your process when making a costume?
I pick a costume based on my abilities and Frankenstein the whole thing. Every piece is planned — what materials, what process. It's all trial and error.

What kind of tech and equipment do you use?
I'm fairly low tech. You can make anything out of EVA foam.

Any favorite pieces?
I built a robot on top of a modded R/C car in a day. It

Natasha aka Bindi Smalls

Tim Winn

Tim Winn

Tim Winn

Natasha aka Bindi Smalls

Tim Winn

wasn't perfect, but I was on a serious time crunch, as well as recovering from surgery.

Advice for aspiring cosplayers?
Keep failing. Embrace it, that's part of the fun of it!

Holly Conrad

hollyconrad.com

I was first interested in cosplay when I was around 5 years old. My favorite costume was a green pillow I taped around my back — a Koopa from *Super Mario Brothers*. After that, I got into going to the Renaissance Faire as a Tiefling from *D&D*, and the rest is history.

What's your process when making a costume?
I love to illustrate, so I take an artist's approach. I like to do my own spin on them, from using interesting, unique materials to aging costumes in the dirt in my backyard. I love being messy when the costume requires it.

What kind of tech and equipment do you use?
I use a lot of molding and casting; a lot of resin and cloth. I've made a few costumes from *Animal Crossing*. I had to challenge myself and do a lot of sewing on top of the foam fabrication. I like to stick to foam and clay. I've been learning to wet felt, which is super exciting and interesting. I've also learned to felt my own witch hats.

Any favorite pieces?
Those that I have a real emotional connection with. I loved being Commander

Shepard because she was such a strong character. My love of storytelling and characters is what drives me to cosplay.

What has been your favorite cosplay moment?
Meeting other people who love the things you do. When I made a costume of the Lady of Pain from my favorite *D&D* setting, one of the creators commented on my blog that I'd done the character justice. It has nothing to do with the likes, and everything to do with the spirit of why we create.

Advice for aspiring cosplayers?
Be true to yourself, ignore the drama that a large community comes with. Make anything. As long as you have passion and drive and kindness, people will come to you and love your work.

Natasha aka Bindi Smalls

bindismalls.com

I got into cosplay after having a habit of making extravagant Halloween costumes. I was disappointed that I could only flex my creativity for this once a year, so cosplay was the answer.

What's your process when making a costume?
I 3D model, 3D print, sew, paint, and do leatherwork for all of my costumes. 3D modeling and 3D printing are my favorite to make armor and props.

What has been your favorite cosplay moment?
When I get to cosplay with my friends as characters from the

Bindi Smalls, Boston McConnaughey, Bryan Humphrey - Mad Scientist with a Camera, Marvin Reyes - CyberHead Designs, Adam Grumbo Films - facebook.com/grumbo

Amie
Danielle
Dansby

Amie
Danielle
Dansby

BONNIE BURTON is a San Francisco-based author and journalist who writes about pop culture, crafts, and all things geeky. Her books include *The Star Wars Craft Book*, *Crafting With Feminism*, and more. She writes for CNET and hosts the "Geek DIY" show.

same video game or series. It's best to cosplay with friends.

Advice for aspiring cosplayers?
Take your time. It's okay to push your deadline to get it right.

Tim Winn
facebook.com/Timforthewinn
When I was a kid my mom made the best costumes. I didn't know what cosplay was back then. Only in the last two years have I turned it into a full time job.

What's your process when making a costume?
I'm known for my foam fabrication. I start with research and use 3D modeling to create templates, which I then convert into 2D patterns. I transfer that to the foam and put it all together like a puzzle.

What has been your favorite cosplay moment?
A few years ago, I was asked to play *Halo* with a very sick young boy. It wasn't a huge event, it was just me dressed as a Spartan from *Halo*, meeting this wonderful family and playing a game we both loved. We had a great time. A few weeks later he passed away. That was hard, but the memory was one of the best I have while cosplaying.

What are you making next?
My job is to create the amazing costumes that you see on the Freakinrad YouTube channel, Twitch, and various commercials. I'm working on a tutorial series about how to build the things we make, and to help people get started.

Advice for aspiring cosplayers?
As long as you keep trying you will succeed.

Amie Danielle Dansby
amiedd.com
The first character that had me interested in cosplay was Anna Valerious from *Van Helsing*. I started with every maker's gateway drug: cardboard and foam. I started in 2012, and it brings me joy to introduce others to the chaos of making.

What's your process when making a costume?
I'm a software developer so I practice project planning for work and for cosplay. I use Trello for my programming sprints, personal projects, and cosplay. I storyboard my cosplay projects into categories. I break down every piece think I'll need, document materials I've used, what worked, and what didn't.

What kind of tech and equipment do you use?
I'm working on a Rufio cosplay and I used a laser cutter for some of the bamboo and leather pieces. I've utilized foam, Worbla, woodshop tools, blacksmith forge, 3D printing, molding and life casting, sewing, electronics, servo motors, and Lego. Each cosplay I have learned a new skill.

Any favorite pieces?
I was really proud of my sword scabbard I made for my Ciri cosplay. The base was cereal boxes. You don't have to spend lots of money to make something look great. Another

Amie
Danielle
Dansby

was the Princess Hilda's staff. I designed the parts in Fusion 360, 3D printed the parts and electronics in two days. This was the first time I designed the CAD parts to fit my design, electronic components, and schematics. Previously, if I needed space, I'd make my design in Fusion 360 with a square or area big enough to shove all the wires and parts into to stay hidden.

What has been your favorite cosplay moment?
We lost my mom to breast cancer. My dad filled the role as mom and dad. We have gone to conventions before, but last year he surprised me at a con and dressed up as Mario riding a blow-up Yoshi costume. My dad has always supported me.

Advice for aspiring cosplayers?
Don't be afraid to be a beginner. I've learned to solve problems, and try new things, and along the way I've met some of the most passionate makers from around the world. ◐

Heirloom Tech

The math and magic of muqarnas

Written by Goli Mohammadi

What is muqarnas and why is it so magical? This is a style of architectural ornamentation that dates back to the middle of the 10th century, originating in Iran and North Africa. In essence, muqarnas compositions are 3D manifestations of 2D Islamic geometric patterns, meaning these insanely complex designs were made using nothing more than a humble compass and ruler.

Employed primarily to form smooth visual transitions between straight walls and domed rooms, muqarnas compositions serve cosmetic, not structural, purposes, and can be found in domes, half-domes, cupolas, niches, arches, and squinches (where the upper corners of a square room are filled in so the ceiling appears domed).

Each muqarnas composition has five common qualities:

1. Three-dimensional shape that can readily be flattened into a two-dimensional geometric outline (and begins as such)
2. Depth of design is variable and determined entirely by the maker
3. Simultaneously architectural and ornamental by nature
4. No intrinsic logical or mathematical boundaries, so can be scaled ad infinitum
5. Consists of stacked tiers made up of cells, each of which has a facet and a roof of sorts

How They're Made

There are two main types of muqarnas: North African/ Middle Eastern, featuring vertical, triangular sections placed next to one another, and Iranian, where horizontal tiers are first created and then connected vertically using geometric segments. Because the styles and geometrics of muqarnas compositions can be so varied, it's difficult to translate into a straight how-to, but Islamic geometries scholar Eric Broug's "Practical Introduction to Muqarnas" video (makezine.com/go/ muqarnas) gives a great, illustrative overview of how to make a simple three-tiered Iranian muqarnas model using foamcore, a pair of compasses, a ruler, a pencil, an X-Acto knife, glue, and straight pins to mount it.

GOLI MOHAMMADI is a word nerd, mountain addict, and former senior editor of *Make:*.

Goli Mohammadi, Sydney Palmer

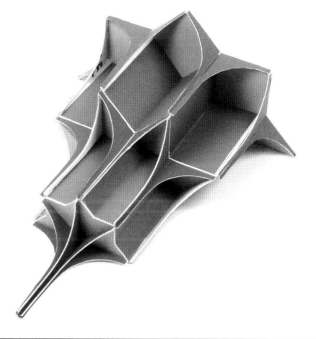

Heirloom

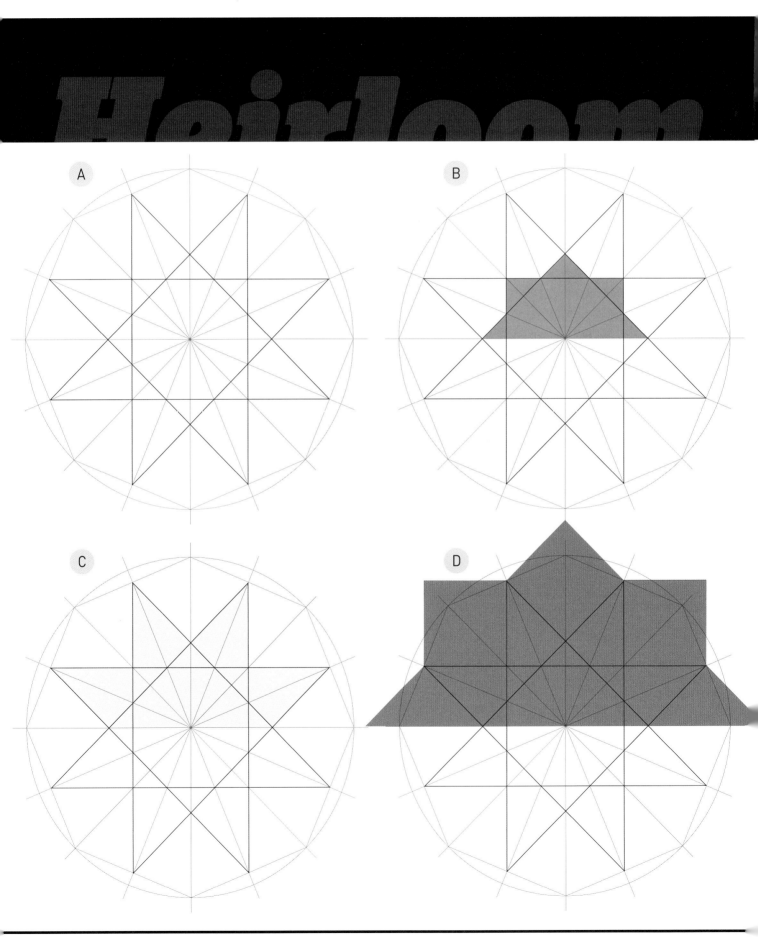

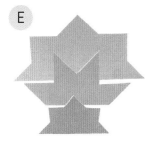

E

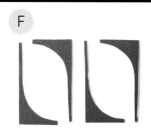

F

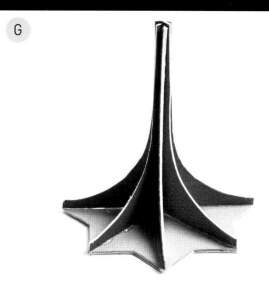

G

Make Your First Muqarnas:

1. Graph a classic 8-pointed star (Figure **A**) using compasses and a ruler. Each tier uses half of the star.

2. The center base design of the star is used for the lowest tier (Figure **B**), and the points of the star are included on the slightly larger middle tier (Figure **C**). The top tier is even larger, including the shape made by adding lines to connect the outer star points (Figure **D**). Cut the three tiers out using an X-Acto knife (Figure **E**).

3. Next, design and cut out the long, thin cardboard elements that connect the tiers (Figure **F**). The length, curve, and geometry of these are completely up to the maker. Once decided, the elements are all uniform in the composition.

4. The connecting elements are then glued on to each tier (Figure **G**) and colored sections of cardboard are cut, folded, and glued along the star pattern (Figure **H**) to add to the composition and make it more solid. The three tiers are then attached together vertically (Figure **I**).

The 2D Behind the 3D
Japanese art scholar Shiro Takahashi has painstakingly cataloged structures across the globe bearing notable muqarnas compositions, mapped by location, and has created readily available simple line drawings for 1061 of the entries. See more at makezine.com/go/shiro1000.

Infinite Material Possibilities

The beauty of such intricate designs based on classic geometries is that they're infinitely scalable in a vast array of materials. No picture can capture how amazing it is to be physically standing under a giant muqarnas composition. I had the pleasure of taking many photos of these thanks to trips visiting family in Iran. Each material inspires a different effect.

The most surreal, perhaps, are the mirror-encrusted muqarnas (Figure **J**), most often found in mosques and shrines. Even though you're surrounded by mirrors, the angles and small size of the individual pieces make it so you never see a reflection, only brightness and light. ✪

H

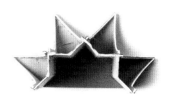

I

J

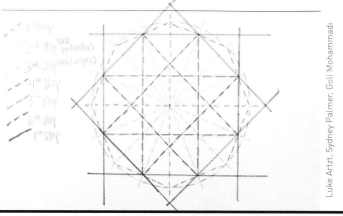

Luke Artzt, Sydney Palmer, Goli Mohammadi

Diamond in the Rough

Rebel Nell's upcycled graffiti jewelry offers women career opportunity and financial independence

Amy Peterson is the co-founder of Rebel Nell (rebelnell.com), a social enterprise in Detroit that fabricates distinctive rings, necklaces, and cufflinks from scavenged chips of graffiti. Rebel Nell uses jewelry making as a means to employ, educate, and empower women. Funded by jewelry sales, they're able to offer financial literacy classes and other resources to impoverished women in local shelters trying to transition to more independent lives.

How'd you start Rebel Nell?

I moved to Detroit and lived next to a shelter. I often talked with the women there, and was inspired by who they are, what they went through, how they wanted to make life changes not only for themselves, but for their families. I wanted to help.

We reversed the usual [startup] process — we wanted to provide jobs and education for women, and we needed to come up with a product to support the classes and resources.

And how did you come up with the product?

[Co-founder] Diana Russell and I love Detroit. We wanted our product to be Detroit-centric. One day I was out running and saw graffiti that had fallen on the ground. I took it home and started playing around with it and got the idea for jewelry. Diana and I spent about four months prototyping. I took some jewelry making classes in school. Diana had taken silversmithing classes.

What's the process?

How we process the graffiti to reveal all the colors is a trade secret. Each piece is made up of many different layers, which we're able to expose. We plate it in silver and then put a resin on top to protect it and bring out its shine. We stamp our logo on the back, loop a silver chain through it, and out the door it goes.

What do you look for in your hiring process?

We look for women who want to change their situation. Do they have a willingness to learn? Do they work well with others? We teach them everything else. When we were designing the jewelry making process, we wanted it to be teachable. We also wanted to build in opportunity for creative input, which serves as a major source of empowerment. They pick whatever colored patterns and shapes speak to them. Each piece is one of a kind, not only because of the craft action and graffiti, but because there's an individual who made it.

How many people do you have on staff?

There are six: four making jewelry, and two working on marketing and sales. Rebel Nell is funded entirely through sales. It takes a lot of hard work and hustle to inform buyers about our product. Our goal is to increase our revenue every year so we can hire more women. In order to do that, we have to come up with unique and innovative ways to share our story online, through grassroots efforts and on social media. Social media and online advertising are incredibly important for reaching younger people — and it's really important to capture them!

How long did it take to get Rebel Nell up and running?

We started in March 2013 with the concept and came up with our product a couple months later. We sold our first pieces in September, and by December we found a spot to work. After about a year, we hired our first three women, with limited hours and limited pay. But they loved it and stuck with us, and fortunately we were able to provide more hours and increase their pay within six months. Now we're in about 35 stores nationwide, and we're constantly growing.

Starting a business is hard enough. What was it like launching a startup with a social mission?

The social mission is what's most important. You have to find a way to balance the mission with the business — for us, that was how to balance the education with production. Obviously if we don't have sales, we can't help the women that we serve.

Sometimes we're heavier on one side than we are on the other, depending on the time of the year and the seasons. But we always reflect back on our mission, to make sure that we're doing what we set out to do.

As a lawyer, do you have any advice to makers who are thinking about setting up their own nonprofit?

It really depends on their business model.

We're set up as an L3C — a limited low-profit liability company, which is an option available in about a dozen states. We've also started a nonprofit arm to support our educational classes.

I suggest to speak to people who will give you free advice, through small business community groups or organizations. It's always great to get the advice of an attorney or accountant too. The first year or so will be challenging, but it's worth it to see the long-term impact on the lives of others.

Any advice for merging skills with a social mission?

You really have to be passionate about the mission. It adds a totally different dynamic. For instance, we're not working with trained artisans and jewelry makers, and we understand that. The jewelry is important, but our focus is on helping individuals trying to get back on their feet. That's a very different approach to business. But I encourage people to do it, because you really can make a difference. And not just in individuals' lives, but their families'. ◗

Read the full interview at makezine.com/go/rebel-nell.

Jacob Lewkow, Sebastian Sullen

DC DENISON is the co-editor of the *Maker Pro Newsletter*, which covers the intersection of makers and business, and is the senior editor, technology at Acquia.

Amy Peterson

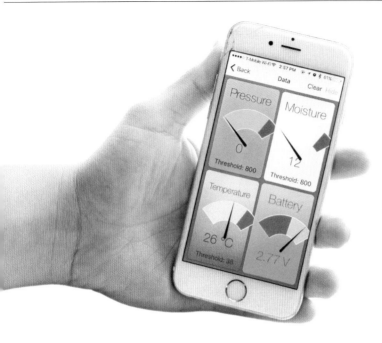

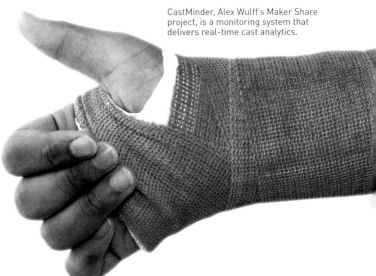

CastMinder, Alex Wulff's Maker Share project, is a monitoring system that delivers real-time cast analytics.

Story Tellers

Written by Matthew A. Dalton

Contribute your projects and collaborate on solutions in the Maker Share online community

There's a story behind every project. Each is a reflection of its creator's life experience and imagination and mindset; understanding the "why" driving a project can be as powerful and interesting as observing the final product itself.

Maker Share exists to help makers tell those stories. This new online community, launched by *Make:* in partnership with Intel, is, at its heart, a story-telling platform: share your ideas, your successes, your failures, and your inspiration. Having shared something fundamental about yourself as a maker, you can connect via Maker Share with potential collaborators around the world.

The platform is built around the model of show & tell, the same model that inspired the first Maker Faire and its tagline,

"The Greatest Show (& Tell) on Earth." The desire to share is a big reason people come to Maker Faire, and Maker Share is built to quench that thirst 365 days a year.

Makers love challenges and solving problems. The necessity of resolving certain issues inspire missions, opportunities to unite around a single problem set and use the talents of the maker community to improve the world we all live in.

Maker Share launched with two missions. Prompted by an appeal from a parent, The Malia Project is focusing on helping Malia, a young girl who has cerebral palsy, communicate more effectively. The Making @ School mission awards $1,000 to a school makerspace in the name of the winning middle or high school student.

While both of these missions are closing shortly, we're excited to announce that our next mission explores projects in the home. Many makers use their homes as large-scale experiments that could improve everyone's living spaces, as evidenced by some of the projects in this issue. Visit makershare.com/missions for the mission details.

Wherever you make whatever you make, you too have a story. Come share it today on Maker Share. ✇

MATTHEW A. DALTON's research into tutorial design, inspired by his involvement with *Make:* and the maker community since attending the first Maker Faire in 2006, earned him a Master of Arts from Simon Fraser University.

Hep Svadja, iStock.com/User9236883_407

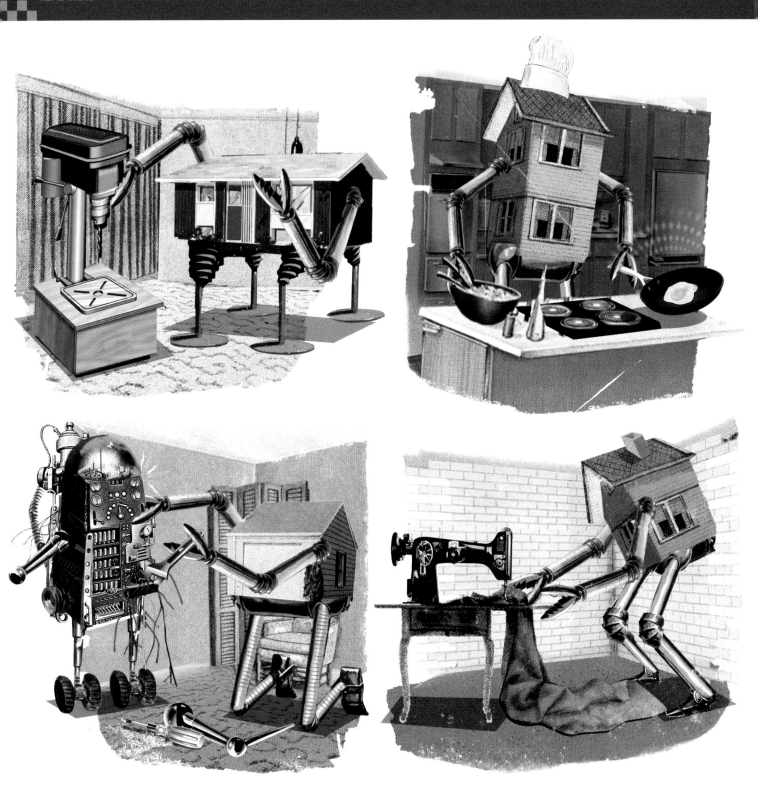

Inside and out, homes are the best medium for people to express
who they are — and no one does that better than makers

~ ILLUSTRATED BY MATTHEW BILLINGTON ~

HOME HACKS

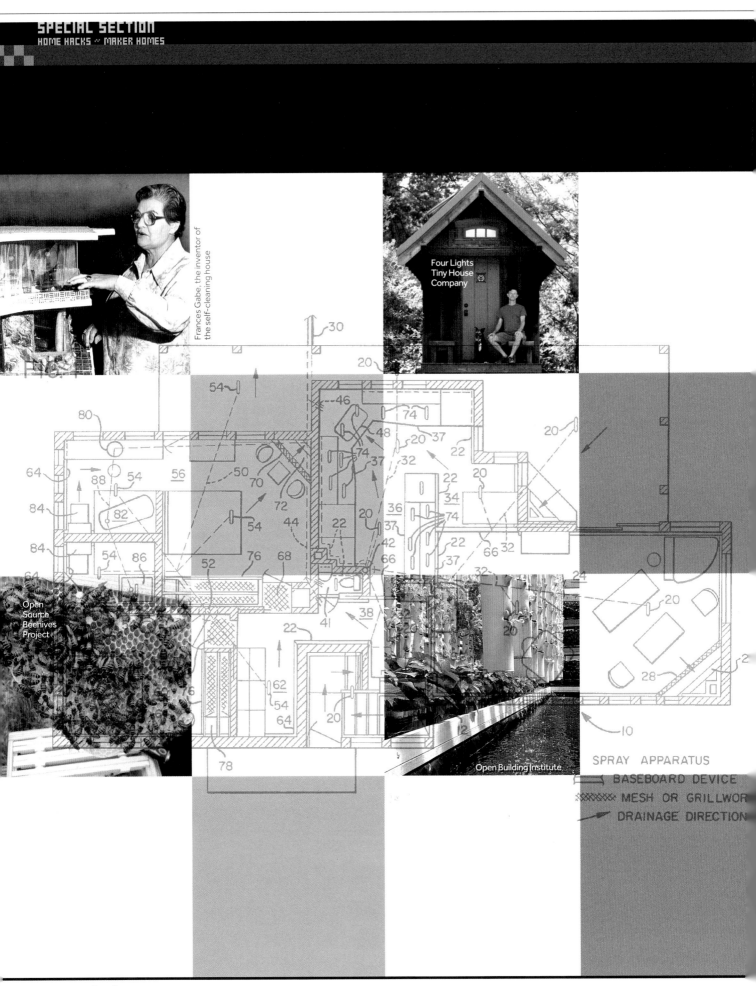

Frances Gabe, the inventor of the self-cleaning house

Four Lights Tiny House Company

Open Source Beehives Project

Open Building Institute

SPRAY APPARATUS
BASEBOARD DEVICE
MESH OR GRILLWOR
DRAINAGE DIRECTION

Jeff DeBoer washroom

HOME-MAKING AND HOUSELIVES

Innovative projects from around the community show how we're reinventing our living spaces
Written by Julia Skott

Aquapioneers

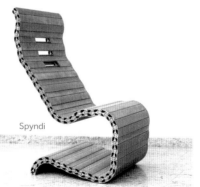

Spyndi

Home is where the hardware is. Some say it is where the Wi-Fi connects automatically, but that goes so beyond without saying in a maker's house. Rather, it is where half the things you see are more than meets the eye, where your light bulb might start correcting you when you talk to yourself, and quite possibly where you might find some sawdust in your morning granola, but roughage is roughage,

right? (The same is not true for yarn — though the fiber pun is right there — or circuitry work. Be careful. Putter responsibly.)

Regardless of your house's size and cleverness, we've all had the dream of being able to just hose everything down. That or a targeted set of very controlled fires. One day, maybe we'll all have our own version of the self-cleaning house

JULIA SKOTT
is a journalist, writer, podcaster, and potter who also knits and tries to keep plants alive. She tweets in Swedish, English, and bad puns as @juliaskott.

FIG. 26

FIG. 25

Art Krumsee, Spyndi, Hep Svadja, Caleb Kraft, Kelly Hofer

that inventor Frances Gabe, who recently passed away, patented in 1984. It functioned basically like a dishwasher, *sans* rotating blades in the middle; splashing, rinsing, and then blowing dry everything not protected by bespoke covers and boxes — including the actual dishes, which would sit out on special racks. (Apparently, the water then took a detour through the doghouse on the way outside, to get the pooch clean as a bonus. Why not.) Until that day:

IT'S WHAT'S ON THE INSIDE

No matter how tiny or smart, you need to have *things* in your house. Decoration? Storage? Seating? Maybe you're still in your boring big, dumb house and want to do what you can with what you have. (Yes, we all want a flip-up wall bed that's also a sauna, a karaoke machine, and does the laundry.)

Opendesk: This London company-cum-platform is a sort of conglomeration of, or maybe dating service for, designers, professional and hobby makers, and people who just want to plonk down money for great and simple furniture. No matter where you are, you can get schematics for a clever and simple desk, say, and then find someone local to you to make it if you're not the jigsaw type yourself. Consumer dollars are less diluted before they reach both designers and carpenters, and your work surface can be just what you want it to be. *opendesk.cc*

Spyndi: Somewhere in the overlap of human vertebrae, caterpillar tank tread, and click-together toys, you find this Lithuanian furniture company, which successfully crowdfunded its hyper-engineered but open-ended chair last year and promises

to deliver to backers some time right about now. Flexible in both senses of the word, the design slots joints of wood together to let you mold a long flat strip into almost whatever you want. For when you maybe don't have time to saw and sand yourself, but still want to get to play around with the shape and function of your chair. Or foot rest. Or ovoid lounging surface. *spyndi.com*

Ply90: Assembling bespoke furniture of any type can be a challenge. The brushed aluminum Ply90 connectors let you piece together almost any configuration of flat material to build sofas, shelves, tables, and more. They're not cheap, but their good looks and versatility make up for that. *ply90.com*

Maker Living Room: *Make:* editor Caleb Kraft and his family came up with a unique

3

1. Opendesk
2. Spyndi
3. Ply90
4. Caleb Kraft's Living Room
5. Jeff DeBoer's washroom
6. Four Lights Tiny House Company

4

5

6

spin on the traditional living room by turning theirs into a makerspace. They replaced the TV entertainment center with a workbench/toolbox/peg board combo, complete with an installed tablet to watch how-to videos while working. Out went the coffee table, and in went a pair of matching worktables with under-bench storage, meaning each family member had their own space for stowing long-term projects. A set of baker's shelves completed the transformation, to hold bulkier tools and supplies.

Rocketing Restroom: As for the most important seat in the house?

Jeff de Boer, of The Little Giant Rocket Company in Calgary, Canada, decided that since shop bathrooms are often grimy anyway, he would turn his into a purposefully grimy one, one that would turn a visit to the

privy into something much more fun.

"I thought as we are a rocket company, it should be a bathroom that looks like it is shared by the lowest ranking crew members," he says.

The project was started during a time when work was slow and everything was over budget, and de Boer was wondering if the whole thing was going to turn out to be a mistake. He decided to go for it and, as he puts it, fail big — which in this case meant, make the bathroom. And as it turns out, 20 years later, the studio is still a success.

"Every time I feel fear, I just go to the bathroom and I have a reminder that taking chances can bring great rewards and joy."

LITTLE BOXES ON THE HILLSIDE

The tiny house movement, and related re-purposing of other structures for living,

seems like a no-brainer match for a maker. You can create your own house more or less from scratch, full of clever solutions and interesting multipurpose assemblies, and you don't need an army of Amish to do it. (Make sure to test a small dance party to get measurements that'll make you happy.)

Four Lights Tiny House Company: Jay Shafer has everything you need — from housing plans for the exterior shells in several styles, to builds for interior components created for maximum small space usage. The Four Lights Tiny Houses are designed on modular interiors, so the bathroom, kitchen, and staircase loft beds can be arranged in several different configurations, and they can be constructed on any base, from trailer beds to stilts or permanent platforms. The step by step

PONDERING A PROTOTYPE

Casa Jasmina in Turin, Italy explores the meaning of the maker home of the future

Written by Bruce Sterling

Massimo Banzi says that you shouldn't work on an open source project unless you're solving your own problems. Since Jasmina and I are a married Serb and a Texan, and we live together in Italy, we've got some problematic issues with making a home.

Casa Jasmina — (the house was Jasmina's idea, and Banzi named it after her) — is our prototype house of the future. It's upstairs from the sprawling Torino Fab Lab. Casa Jasmina is where Jasmina and I, along with our Turinese friends, experiment with domestic maker culture and the open source Internet of Things.

MODEL HOME

I was a *Make:* columnist in the mag's early days, so when Jasmina and I first arrived in Turin in 2007 we fell in with Italy's first Fab Lab. Like a lot of European makerspaces, the Torino Fab Lab exists inside a big derelict factory. In 2015, we launched Casa Jasmina; it took our collaborators and us six months to loft-out a space fit to live in. We transformed musty, rusty old factory offices into a prototype Italian family home, where most objects and services are made by open source, maker-style methods.

An Arduino office also exists inside this factory, so we specialize in light electronics: tunable lights, a smart thermostat, a keypad lock, and an automatically irrigated garden. Toolbox Co-Working, the local design office space, finds the house useful. The Internet of Women Things group meets inside Casa Jasmina, where women in the maker scene discuss ideas for a good and ethical home in an era of powerful, intrusive IoT home automation.

So this prototype "house" has a lot going on: It's a guest residence, a test-bed for open-source hardware projects, and a public showplace. We use it to throw workshops, art events, and big parties. Whenever we learn something useful in Casa Jasmina we tend to port it across town and install it in our own apartment. We don't work for Casa Jasmina — we're just the founders, the hostess and the curator — but it's been very handy for us.

A house created with maker objects and techniques is an intriguing idea, and Italians adore innovative home design, so it's been a public-relations success. Journalists drop by, politicians, museum curators, even an Italian astronaut have appeared, and they tend to say that it's quite cute.

In contrast to the big Torino Fab Lab downstairs, with its drill presses, laser cutters, and six-axis robot, Casa Jasmina is a cozy, pleasant, even a stylish area. People can cook in there, sleep there, wash, enjoy a glass of prosecco in the flourishing garden. Casa Jasmina even features a kids' room, which, by the nature of kids' products, is its

1. Casa Jasmina
   under renovation
2. Casa Jasmina
   design meeting
3. Terrace Yard,
   before renovation
4. Lorenzo Romagnoli
   and Alessandro
   Squatrito of Casa
   Jasmina getting
   their hands dirty
   in the new terrace
   garden
5. Bricks and debris in
   what becomes the
   workspace area
6. The finished main
   hall and workspace
7. The new Casa
   Jasmina kitchen
8. The final floor plan

BRUCE STERLING is a novelist, world traveller and former *Make:* magazine columnist.

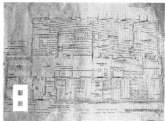

most genuinely "futuristic" aspect.

Now, though, Casa Jasmina is over two years old. As a building project, it's been completed. We've learned that maker culture can accomplish surprising things. However, as a design lifestyle, "making" is missing important cultural elements.

SOCIAL EXPERIMENT

The maker scene excels at composing things that can be described with Instructables, recipes, and algorithms, and can then be simply constructed from flat-pack elements with numerically controlled machines. We understood that from the beginning, and wanted to get intimate with that situation, and we did. As a married couple we now excel at boldly placing our technical needs front and center in our way of life. We'll cheerfully drill holes through most anything we possess. We commonly lash things together with Bluetooth and zip ties.

But we have no maker kitchen appliances. White-goods machines are out of reach of the maker ecosystem, being too legally regulated, too complex, too hazardous to users. Plywood router-cut furniture is indeed easy to assemble, but also falls apart too easily. Plastic 3D-printed connectors are versatile, but weak and wobbly.

Maker culture is internet-centric. It has attitude problems with cherished heirlooms, local craft traditions, and unique local materials.

A hacker scene is all about Do-It-Together, but this clubby approach lacks respect for family members who don't hack, and can't or don't want to learn. Grandma in her wheelchair, baby in her crib, they get sidelined, treated as burdens rather than honored family members. Guests who confront weird open source interfaces have to stare and scratch their heads. And the Internet of Things has serious spy and security issues: it imports cyberwar and cybercrime straight into the bedroom and bathroom. Such is the Casa Jasmina real-life.

People tend to call Casa Jasmina an art installation because, although it would be technically possible to make a thousand Casa Jasminas all over this planet, so far there is only one in the world. Why? Well, as an open source project goes, yes, we did solve our own problem. As husband and wife, we both bring a lot of foreign baggage to our home. But we're OK with that: We've learned to upgrade, duct-tape, and zip-tie our alien things like nobody's business.

Our next problem, though, is to figure out what this might potentially mean to everybody else. We're thinking your guess is as good as ours.

HUMANITARIAN HOUSING

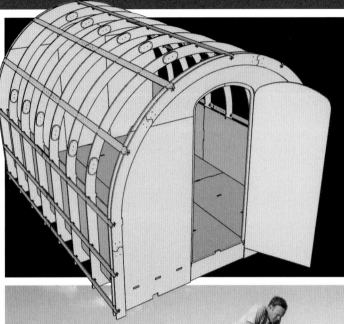

Conceived as transitional emergency dwellings, the flat-pack **Shelter 2.0** design can be easily built from CNC templates with a hand router

Written by Bill Young

BILL YOUNG is a boat carpenter turned CNC evangelist and co-founder of Shelter20.com and 100kGarages.com. Mostly he turns plywood into dust and noise.

Shelter 2.0 in Haiti.

Shelter 2.0 built from construction lumber.

Bill Young

Shelter 2.0 is a flat-packed plywood structure that anyone can build with the assistance of a CNC router. For years my longtime Shelter 2.0 (shelter20.com) collaborator Robert Bridges and I have been shipping Shelter kits all over the world. The Shelter 2.0 project began life as Robert's entry in a contest hosted by the Guggenheim Museum and SketchUp. Inspired by the iconic Quonset hut, it incorporated a barreled ceiling that we had worked on as we admired the strength and aesthetic appeal of those sorts of curved shapes ... a ShopBot can cut a curve as easily as a straight line.

Although originally conceived as transitional housing in emergency situations, the Shelter 2.0 design has been used as housing for the homeless, as team-building projects in schools, and is in the planning stages to provide an option for refugee housing through the Global Humanitarian Lab (globalhumanitarianlab. org). It was designed for efficient CNC cutting; we make the CNC files available online (github.com/ wlyoung/Shelter20) so that anyone can cut them, but it requires a CNC machine. We can easily cut and ship them anywhere in the world, or we can ship CNC machines so that Shelters can be cut on site.

Unfortunately both of these options have some problems. Setting up a shop can be expensive, and then teaching someone to run the equipment on short notice is problematic at best, especially when a disaster hits. Cutting and shipping structures is slow and costly, or requires stockpiling, which ties up money.

DIGITAL MEETS ANALOG
The Shelter 2.0 design requires multiple copies of a small number of pieces. The trick is to duplicate those copies accurately without having to CNC cut every one of them. That's exactly how it worked when things were strictly analog.

CNC cutting the original patterns would give us the accuracy that we struggled with when hand-making templates. Once these patterns are created, anyone can duplicate them with easily available hand power tools. A hand router with a flush trim bit works well for cutting out pieces, but a plunge router with properly sized templates and a guide bushing also makes it easy to cut holes and mill features that don't go all the way through.

Over the years Robert and I have worked on quite a few innovative CNC-fabricated house prototypes, and in spite of how careful we and the designers had been, once we started to assemble the structure we would often find at least one missing slot or even a whole part that required on-site fabrication. We've never had a CNC machine with us but always seem to have a hand router, and have successfully used this duplication technique to copy a whole part on the job using an existing part as the template, or just add a missing slot, dogbone, or some other feature.

PACKED WITH POTENTIAL
Not every technique works for all cases and this one is no different. There are lots of amazing projects and designs that are only possible using digital fabrication tools from start to finish, but there are others where using a similar combination of digital and analog techniques makes a lot of sense. Although we originally became interested in this technique for fabricating Shelter 2.0 in sub-optimal conditions, we're exploring fabricating all sorts of products where a CNC machine might not be available.

I'm still incredibly passionate about the potential that digital fabrication, especially CNC routing, has for revolutionizing the way that much of the things in our lives are made. If you already have access to a CNC machine, put it to work! But if not maybe this digital/analog hybrid system is worth a look. ◎

THIS WORKS BEST WHEN:
» There's a limited number of unique parts in a design, but several copies of each are needed.
» The parts require mostly 2D outline cutting (although some 3D features like pockets are certainly possible).
» There is an excess of available labor. The process can branch and grow as extra labor becomes available.
» There's poor security or electrical infrastructure. All the tools needed for this system are small, can be powered by a small generator or solar panels, and can be locked in a closet or secure toolbox.
» The design is somewhat "material agnostic," allowing the use of a wide range of materials.
» You don't need absolute precision, or don't mind a couple of screw holes on your parts.

WHAT ARE THE BENEFITS ON A LARGE SCALE?
This works really well occasionally for duplicating a single missing part or feature, but does it make sense as a manufacturing technique? Here's why we think it does:

» The files are digital, so patterns can be CNC cut locally, or shipped from elsewhere. A set of patterns is a much smaller package to transport, and only needs to be done once.
» Most of the knowledge and all of the precision is built into the CNC-cut pattern. Techniques for duplicating the patterns with traditional hand power tools (drills, jigsaws, and hand-held routers) can be learned quickly.
» It can expand. You can increase the number of parts you can produce by simply duplicating the templates on-site and adding more people and hand tools.
» It's easy to "backup." A copy of the template can be created and used as the working set, saving the original.
» It can work with materials that are hard to CNC cut.

Fabricating table legs using this digital/analog technique.

RUSSELL GRAVES is a 35-year-old husband and father with a habit of taking things apart, documenting the process, and then repairing or improving them.

TIME REQUIRED:
›› 3+ Weeks

COST:
›› $16,000 in initial build, plus another $1,000 to $1,500 in eventual upgrades.

MATERIALS
›› **Pre-built Tuff Shed**
›› **Rock wool insulation**
›› **Blue foam board insulation, 2"**
›› **Great Stuff Foam**
›› **Plywood, ½" and ¼"** for the walls and ceiling
›› **SolarWorld Sunmodule panel 285W (10)**
›› **6/2 UF and 10/2 UF** for panel wiring
›› **Assorted lumber and bolts** for panel mounts
›› **Charge controller** MidNite Classic 200 MPPT
›› **Deep cycle flooded lead acid batteries (8)** Trojan T105-RE 6V
›› **Inverter/charger** Aims Power 2000W 48V
›› **Circuit breakers and combiner boxes** for panels
›› **Air conditioner and heat pump** Frigidaire FFRH0822R1
›› **Plywood, ⅝" and shelf braces**
›› **Wireless radios and access point** Mikrotik mikrotik.com

TOOLS
›› **Circular saw**
›› **Reciprocating saw**
›› **Drill**
›› **Impact driver**
›› **Wire crimpers**
›› **DC lug crimpers**
›› **Screwdrivers**
›› **Inverter**
›› **Generator**

OFF-GRID OFFICE

How I customized a Tuff Shed into a **solar-powered workspace**
Written and Photographed by Russell Graves

A year and a half ago, I moved from the greater Seattle metro area to a rural property in southwest Idaho — and I needed a place to work, so I built myself a solar powered office out of a Tuff Shed! I've worked from home in the past, and knew that I need a separate workspace to maintain isolation between my work and the rest of my life. And also because I work with lithium batteries, small electronics, soldering irons, spot welders, power supplies, and other dangerous equipment that isn't exactly kid-safe.

BUILDING THE OFFICE
It took me three weeks of full-time work to go from bare ground to a functional office. During this time, I built a foundation (I got the shed at a discount as a fully built unit, which required a smooth and level surface), helped slide the shed into position, foam-insulated gaps around the windows and doors, stuffed the walls and ceiling with rock wool insulation, put a layer of foam board over the insulation, installed plywood walls, cut a huge hole in a wall for an air

conditioner, installed a ton of shelves, built myself a lab bench on one end, hooked up a large battery bank in a box outside, built myself solar panel mounts, installed the charge controller and inverter, and ran wireless internet to the house — it was quite the project!

OFF-GRID POWER

My office is solar powered because my property is a thin layer of dirt on top of a base of basalt. The house is connected to the power grid, but trenching around here usually involves explosives. I've been interested in off-grid solar power for years, and this project offered me an amazing opportunity to build such a system, deal with it on a daily basis, and learn a ton about actually running an off-grid power system all year long.

Energy comes from 10 SolarWorld 285-watt Sunmodule panels. I have 8 of these mounted on south-facing wooden frames, and the remaining two hinged on my east wall. I call those two my "morning panels," and they very effectively capture the sun coming over the horizon before the main panels are producing much power. If needed, I can swing these around to the southwest to capture extra sun during the afternoon and evening.

Power from the panels flows through a MidNite Classic 200 MPPT charge controller into 8 Trojan T105-RE flooded lead acid batteries, then into an Aims Power 2000W inverter/charger (6000W peak).

INTERIOR

Inside my office, the computers sit on a wrap-around desk I've had for over a decade. A Raspberry Pi 3 runs constantly to monitor the power system, a large desktop runs Folding@home (folding.stanford.edu) and BOINC (boinc.berkeley.edu) if I have excess power, a power-sipping iMac handles general desktop duties, and a few laptops round out my computing needs.

On the other side, a 7.5'×2' lab bench serves as my working area for building and tearing down battery packs, designing and analyzing small electronics, and whatever else I feel like doing. I've got a soldering station, spot welder, oscilloscope, bench voltmeter, and various other tools easily available.

WINTER

Winter in Idaho is cold and snowy — which isn't great for a solar-powered office. A small inverter-based generator provides power for battery charging on cloudy days, and a ventless propane heater brings the interior temperature up on chilly mornings. If it's a sunny day, I have no trouble — the electric heating elements in my air conditioner and a small under-desk heater work beautifully. The rock wool and foam board make for a well-insulated shell that keeps me warm all winter long. The batteries get cold sitting outside, but lead acid doesn't freeze unless it's extremely cold or the batteries are deeply discharged.

INTERNET AND WORK

I work remotely from my office, so a pair of rural wireless internet connections (one on the house, one on the office) keeps me connected to the world. The house and office are linked with Mikrotik wireless radios, and I can run the entire property off either connection.

My work qualifies as "deep work" — I focus on deep technical tasks for long periods, and having a distraction-free environment I can customize to my needs is absolutely incredible. With a cordless drill, I can modify my office whenever I feel like it, and there's nobody tapping me on the shoulder to ask if I feel like lunch right now. I've worked in many different environments, and this is, by far, my absolute favorite. I cannot say enough good things about having your own isolated workspace for deep technical work!

GOING FORWARD

An office like this is a perpetual project. As I find things that don't work well, or that I want to change, I can simply change them. An impact driver and a box of screws, combined with the plywood walls, means I can mount things wherever I want. I'm also increasing the insulation before next winter with some under-floor foam and some window plugs.

I'm very happy with how my office turned out. More space would always be nice, but the smaller interior space is easier to heat and cool, and kept the finishing costs lower. A team of people would certainly make the initial setup quicker! ◐

The Tuff Shed being delivered on a truck.

Interior insulation.

Solar panels angled for maximum sun.

Banks of 8 Trojan batteries to store power.

Computer monitors on a wraparound desk.

During the winter snow can block panels.

You can find more details about this project, as well as Graves' other work, on his blog: syonyk.blogspot.com

ALTERNATIVE ABODES

A variety of ways to DIY your living space from the ground up

Written by Goli Mohammadi

The modern maker's fixation with reuse, cost-efficiency, and conservation has spawned much outside-the-box thinking when it comes to dwellings. If cookie-cutter suburban tract homes are on one end of the spectrum, these alt. homes are on the other. Whether through use of clever materials or novel forms, these cozy habitats are simultaneously kind on the planet and wallet. Your biggest inhibitor, though, may be getting a permit.

GOLI MOHAMMADI is a word nerd, mountain addict, and former senior editor of *Make:*.

CLEVER MATERIALS:

1. PALLETS

Pallets are widely used, readily available, and often free. There's a whole movement of folks using pallets to make furniture, doghouses, sheds, and full (but usually small) homes. Be forewarned that pallet wood is typically either heat-treated (safe) or treated with methyl bromide (toxic). The International Plant Protection Convention requires pallets to be stamped with a two-letter country of origin code, a unique assigned number, followed by either HT or MB denoting how it was treated.

To err on the safe side, if there's no stamp, you're better off not risking chemical exposure. Pallets are usually disassembled into planks, or used as-is and supplemented with insulating material. The roofs are often made with a lighter material like tin or corrugated plastic. You can make a 16'×16' structure with about 100 pallets.

■ **How-to:** makezine.com/go/pallet-house

2. SHIPPING CONTAINERS

Almost everything we buy overseas is shipped in a corrugated steel shipping container. Because America exports so much less than we import, and the cost of shipping an empty container back outweighs its worth, we literally have piles of them sitting around at our ports.

Ranging in price from roughly $800 to $5,000, depending on size and condition, shipping containers offer a massive building block that is ripe for stacking, perfect for customizing, and easily welded to cut down on building time and cost. What's more, because they were made to withstand the harshest weather in transport, they're more durable and resistant to the elements, including fire and termites.

■ **How-to:** residentialshippingcontainerprimer.com

3. EARTHSHIPS

Developed by architect Michael Reynolds in the '70s, Earthships use locally abundant natural materials and trash to create sturdy, off-grid structures. Stacked tires packed tightly with dirt become free giant bricks. Layering bottles and cans into cement minimizes the cement needed.

Intended to eliminate reliance on public utilities and fossil fuels, Earthships are designed with built-in systems such as rainwater catchment, passive solar, and greenhouses to collect and heat water, produce food, and heat/cool the house. They're intentionally simple so anyone can make one. Reynolds convinced New Mexico to give him a couple of acres to build an experimental Earthship village, free from codes and regulations. Plans and workshops are available through Reynolds' Earthship Biotecture.

■ **How-to:** earthship.com

4. EARTH BAGS

One of the chief complaints about Earthship homes is the alleged off-gassing of the main building material, tires, particularly in hot climates (though this issue is mitigated when the tires are covered in cement or adobe). An alternative building block is the "earth bag," quite literally polypropylene bags filled with dirt and rocks, then sewn shut and stacked neatly with barbed wire in between layers to secure and add tensile strength. Earth bag structures are typically covered in adobe or a similar material.

■ **How-to:** makezine.com/go/earthbags

5. STRAW BALE AND COB

Cob is a versatile and ancient building material made from a mixture of subsoil (the soil beneath topsoil), water, something fibrous (such as straw), and sometimes clay, sand, or lime. Used since prehistoric times, it's fireproof, inexpensive, and can usually be locally sourced. And while you may be picturing a structure that looks like it came out of *The Hobbit*, modern cob buildings can be anything but, taking on more conventional forms.

Oftentimes, hybrid structures are made, combining cob construction with straw bale building, which employs bales of straw as big building blocks. A hybrid house may use straw bale on the north side, where you want the most insulation, and cob on the south side to absorb heat and distribute it into the house. Also, cob is easier to put windows in and make curves, while straw bales make for a quick and easy build.

■ **How-to (straw bale):** simple-living-today.com/straw-bale-house.html
■ **How-to (cob):** diynatural.com/cob-house-construction

NOVEL FORMS:

6. TINY HOME ON WHEELS

The tiny home movement is all the rage, with builders using clever storage and dual-use spaces in these typically 400-square-foot or less houses. Now that movement has sprung wings, or rather wheels. These tiny homes are on the move, and there's no better way to evade property taxes while ensuring the best view.

Canada-based builder Laird Herbert offers inspiration with his signature Leaf House designs (facebook.com/LeafHouseSmallSpaceDesign AndBuild). His Version 3, for instance, is a 215-square-foot home made for cold weather, built with recycled and sustainable materials, that fits on a 20-foot trailer, weighs less than 5,500 pounds, and can fit a family of four.

■ **How-to:** instructables.com/id/Tiny-tiny-house

7. HOUSEBOAT

Houseboats aren't new, but do offer an interesting alternative. There are houseboats and also "floating homes" — the latter is a structure built on a floating platform, while the former must have seaworthy hulls, an engine, a navigational system, and meet U.S. Coast Guard standards.

On a houseboat, you have no yard to maintain, the view can always change, and in some places, they're taxed as private property instead of real estate. There is, however, typically a mooring fee and sometimes homeowner's association fees. Plus, your homeowner's insurance is typically higher than land-based structures, and your home is more exposed to the elements.

■ **How-to:** buildahouseboat.com ◢

Paletten Haus, Angel Schatz, Biodiesel33, Kelly Hart - www.earthbagbuilding.com, Casa Terracota, Laird Herbert, Waqcku

VOICE BOX

Speech recognition for Arduino has arrived — and it could change the meaning of a smart home

Written by Jon Christian

Arduino, let there be light.

Christopher Coté is speaking sternly to a Pixar-style desk lamp.

"Arduino," he says, in the tone of voice you might use to catch Siri's attention, and it chirps to show that it recognized the call sign. "Let there be light." In response, the lamp blinks on, illuminating Coté's face with a warm glow.

UNDER CONTROL

Coté, a researcher at CRT Labs, built the lamp to experiment with MOVI, an Arduino shield designed specifically to provide onboard speech recognition and synthesis. It's notable less for any single breakthrough than for wrangling a handful of existing open source speech tools into a board that makes it dirt-simple to add voice control to an Arduino project. And unlike an off-the-shelf smart speaker like Google Home or Amazon Echo, MOVI does it all locally — nothing gets sent to the cloud, alleviating concerns about privacy and security.

"The driving factor for us was the ability to be completely disconnected," said Bertrand Irissou, one of MOVI's creators. "It's great to have all these devices connected to the cloud, but if the internet goes down, they won't work anymore."

OPEN HOUSE

Watching MOVI in action, it's easy to imagine a future in which smart homes look less like Google's or Amazon's walled gardens and more like Linux: an ecosystem of crowdsourced, customizable systems that provide the functionality you need without giving up control.

Take Steve Quinn, another early adopter who works by day in the British space industry and spends his free time tinkering with open source smart home technology. When he unboxed his MOVI board, he quickly configured it to send commands through an ESP8266, via the MQTT IoT protocol, and into OpenHAB, an open source home automation system he'd already set up in his house. Before

he knew it, he could access his home's lighting and sensor networks by talking to them. Next, he plans to add code to control his television and security cameras.

"It's an outstanding kit," he said. "If the price comes down, I'll have a few more of them."

STILL KICKING

MOVI began its life about two years ago as a Kickstarter campaign by Irissou and a collaborator named Gerald Friedland. Demand continued after they delivered the initial units, so they've kept selling boards online for about $75 each, and they've seen users build everything from a voice-controlled wheelchair to interactive art installations. One of Irissou's favorites is a voice activated Iron Man suit by a cosplayer named Julius Sanchez.

"The guy is not a programmer," Irissou said. "He's a maker with some rudimentary coding skills, and he was able to do things that just flabbergasted us."

Juliann Brown

Smarter Objects

Add IoT functionality to common items with MESH   Written by Sam Brown

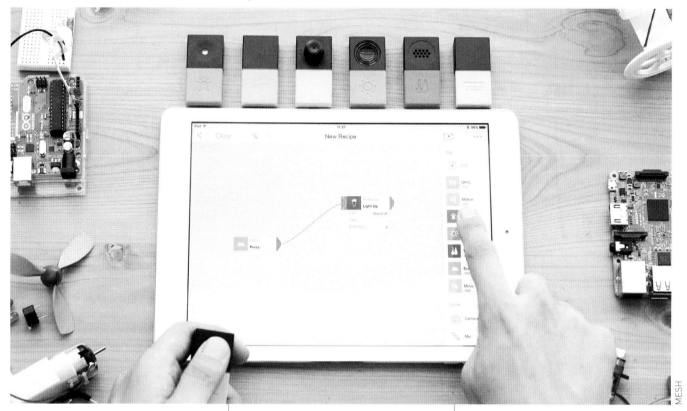

MESH

Compact MESH programmable blocks help you add smarts to your everyday objects without needing to write a line of code or untangle a rat's nest of wires.

Each block contains a single feature, like a motion detector or a button. Instead of writing code to use them, you open up an app on your smartphone or tablet, and draw connections between the blocks showing which controls which.

MESH is made to be tidy and simple. They communicate wirelessly, and you can even run each block unplugged, since they all have built-in batteries, which last for approximately 30 days.

Starter Projects

The simplest way to start is to have a MESH block notify you via your smartphone or tablet when the block is activated. Leave a MESH motion detector by the front door and get a notice when anyone walks by. Place a light sensor inside the cookie jar and get a notification when

light shines in. You can even use MESH blocks to make a DIY voice-activated self-watering plant stand — find the how-to at makezine.com/go/mesh-plant-waterer.

MESH Plays Well With Others

My favorite part of MESH is how readily it works with other smart devices. If you don't have one already, sign up for an IFTTT account; controlling other gizmos and services with MESH sensors will work the same as using your phone. Just open the MESH app, add your smart device, and then draw a line connecting that gizmo to the MESH block that will control it.

Setting up a MESH sensor to control other smart home products, like a Nest thermostat or WeMo coffee maker, is also easy. Put a MESH light sensor in your kitchen or office, and tell your smart coffeepot to turn on when the lights come up. You'll have coffee fresh the moment you're ready for it. Sleep in on the weekends and it won't be brewed until

you're actually up. Come into the office on a Saturday, and you'll be greeted by a brew starting as you turn on the lights.

Beyond Beginners — GPIO and JavaScript

MESH may be simple, but it's not limited to beginners. If you've worked with microcontrollers before, you'll find the usual controller pins on MESH's GPIO block. Through these pins you can interface with countless sensors and actuators. You can even use them to connect to an Arduino or Raspberry Pi, and tie them to the effortless, wireless sensor connections you build with MESH.

MESH's programming options are likewise open ended. If the pre-built programming blocks don't have the feature you need, new programming functions can be added in JavaScript, too.

Any maker who wants fast, clean results can appreciate MESH.

SECRET CABINET LOCK

A hidden capacitive touch switch allows only those in the know to open it

Written by Mark Longley

MARK LONGLEY is the CEO of Xkitz Electronics which he founded in 2010 with his teenaged children as a sort of real world teaching exercise in entrepreneurship with a goal of marketing the electronic hobby projects they enjoyed designing and building.

We wanted to lock up the cabinet where we store medications and other things we keep away from the kids. But adding any type of standard lock would ruin the look of the cabinet. We didn't want to see any visible sign of the locking mechanism.

We decided to use an electric solenoid type of lock that would be triggered by a hidden capacitive touch switch. A piece of conductive copper tape stuck on the inside of the cabinet is monitored by a capacitive touch switch module. Whenever you touch that spot on the outside of the cabinet, the touch switch module detects the touch and energizes the solenoid lock, which releases the door to open.

So now, we have a totally hidden lock on our medicine cabinet that only we adults know how to open (as long as we don't let the kids see how it works)!

1. POSITION AND MOUNT THE SOLENOID LOCK

The solenoid lock is placed inside the cabinet with the flat part of the plunger facing outward so that the striker plate will push it open when the cabinet door closes (Figure Ⓐ). The lock was oriented 90° from what we needed for proper mounting (Figure Ⓑ), so we removed it from its steel housing and attached it to a piece of aluminum angle iron that would let us easily mount it to the inside of the cabinet.

We decided to add two extra "fail safe" wires from the solenoid to be hidden in the adjacent (unlocked) cabinet, just in case the touch sense mechanism fails at some point. With this, we can simply connect a 12V DC voltage to those wires to activate the solenoid and open the door.

2. RUN 12V DC TO THE CABINET

The touch sense module and the solenoid run on 12V DC. We used an old wall wart

Hep Svadja, Mark Longley

TIME REQUIRED:
» 2 Hours

COST:
» $55–$75

MATERIALS
» **Lock-style solenoid, 12VDC** Adafruit #1512 adafruit.com
» **Capacitive touch switch** I used Xkitz Electronics #XCTS-1M xkitz.com
» **Power supply, 12V DC**
» **Copper or aluminum tape**
» **Wire cable, about 20–24 AWG** Buy it by the foot at OpenBuilds openbuildspartstore.com
» **T bracket**
» **Wood screws**

TOOLS
» **Soldering iron**
» **Drill**
» **Scotch tape**

we saved from some discarded device for our supply. We tucked the wall wart into the recessed light compartment above the sink so that the wiring would be hidden (Figure Ⓒ). Just a few holes drilled inside and between the adjacent cabinets allowed us to run the 12V wire into our cabinet to be locked (Figure Ⓓ).

3. BUILD AND POSITION THE TOUCH SENSE PLATE

Two 5" lengths of copper tape are used for the touch sense plate. These should be affixed parallel and close to each other, but not touching. One of them will be the touch sense plate and the other will be the "ground plane," which improves the touch sensitivity. We used Scotch tape to hold the two pieces together prior to affixing them to the inner wall of the cabinet. Solder (or just tape) a 6" length of thin wire to each of the pieces of copper tape (Figure Ⓔ). These wires will connect to the touch switch module and allow the touch detection. Peel off the paper backings and affix the tapes to the inside of the cabinet behind where you want the touch sense area to be.

At first we tried using the small touch plate PCB that came with the touch switch (Figure Ⓕ). But it turned out to be too small to allow reliable touch sensing through the ¾" thickness of the wood. We had to set the sensitivity of the module very high in order for it to detect touch. That worked pretty well, but sometimes it had false activations because the sensitivity was set so high. Switching to the copper tape allowed us to turn down the touch sensitivity and now it's working rock solid. Presumably, the small PCB would work if your cabinet wall is thinner.

4. WIRE THE TOUCH SWITCH MODULE

You now have all the wiring ready to connect to the touch switch module. Pass

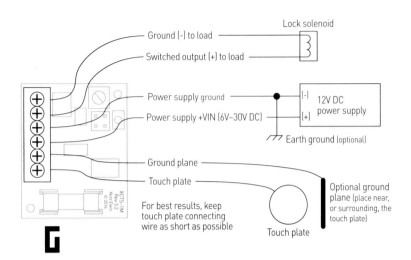

Lock solenoid

- Ground (-) to load
- Switched output (+) to load
- Power supply ground
- Power supply +VIN (6V–30V DC)

[-] 12V DC power supply [+]

Earth ground (optional)

- Ground plane
- Touch plate

For best results, keep touch plate connecting wire as short as possible

Optional ground plane (place near, or surrounding, the touch plate)

Touch plate

G

Connection and Jumper Quick Reference

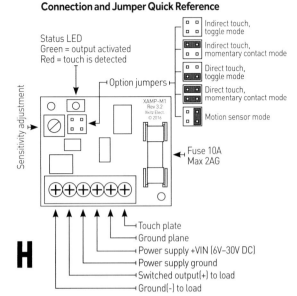

Status LED
Green = output activated
Red = touch is detected

Sensitivity adjustment

Option jumpers

- Indirect touch, toggle mode
- Indirect touch, momentary contact mode
- Direct touch, toggle mode
- Direct touch, momentary contact mode
- Motion sensor mode

XAMP-M1
Rev 3.2
Xkitz Elect.
© 2016

Fuse 10A
Max 2AG

- Touch plate
- Ground plane
- Power supply +VIN (6V–30V DC)
- Power supply ground
- Switched output (+) to load
- Ground (-) to load

H

I

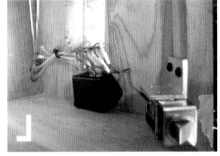

J

K

L

M

N

the wires from the power supply, solenoid, touch plate, and ground plane through the small hole in the side of the touch switch enclosure, and connect them to the small green terminal block as shown in Figures **G** and **I**. Be careful not to wire the power supply backward, it will blow the module!

5. CONFIGURE, TEST, AND ADJUST THE TOUCH SWITCH MODULE

Set the two jumpers on the touch switch module to "Indirect Touch, Momentary Contact Mode," as shown in Figure **H**. Use a small screwdriver to turn the sensitivity pot all the way counter-clockwise to the lowest sensitivity setting and plug in the 12V power supply. You should see the LED on the module blink a few times. Gradually turn the pot clockwise while repeatedly touching the touch sense area until you get to a workable setting.

Important: You must remove your hand from the area between tests, as the module "learns" the ambient capacitance of the touch plate. Keeping your hand on the plate lets it become accustomed to its

capacitance so that it no longer detects any difference when you touch it.

6. MOUNT THE TOUCH SWITCH MODULE

Once the module is adjusted and working, you can mount it to the inside of the cabinet with two small wood screws through the mounting flanges (Figure **J**).

7. POSITION AND MOUNT THE STRIKE PLATE ON THE DOOR

We fabricated a strike plate from a small T bracket (Figure **K**). To find the proper alignment, temporarily put pieces of masking tape on the edge of the cabinet and the door so you can mark the vertical position of the lock and transfer it to the back of the door (Figure **L**). Then measure the horizontal distance and the depth of the solenoid lock. Bend the T bracket to the proper depth (Figure **M**), cut the excess if necessary, and mount it on the door (Figure **N**) to allow it to engage the lock.

You can now keep things safe in your hidden-lock cabinet, and open it with a secret touch. ✪

Juliann Brown, Mark Longley

BUSINESS REPLY MAIL
FIRST-CLASS MAIL PERMIT NO. 865 NORTH HOLLYWOOD, CA

POSTAGE WILL BE PAID BY ADDRESSEE

Make:

PO BOX 17046
NORTH HOLLYWOOD CA 91615-9186

a project by Make: | intel
Maker
Share
make on!
Introducing Maker Share: where objects become innovation
COME SHOW & TELL WHAT YOU CAN DO
maker portfolios | community missions | learning center
MAKERSHARE.COM
Becca Henry Photography

PIRATE FINDER

Keep an eye on your network's sign-ons to spot unwelcome intrusions with this oversized counter

Written by Alasdair Allan

Hep Svadja, Luke Arztz

ALASDAIR ALLAN is a scientist, author, hacker, and journalist. He has mesh-networked the Moscone Center, caused a U.S. Senate hearing, and contributed to the detection of what was — at the time — the most distant object yet discovered.

We all want to protect our internet from unwanted users who may be hogging bandwidth. This project will show you, with huge bright numbers, how many devices are connected to your system in real time. If you notice an unusual increase, you may have uninvited guests, and should look into changing passwords and undertaking other security measures.

Here are the essentials to assembling this build. Find detailed steps at makezine.com/go/network-counter.

1. CONFIGURE RASPI

» Install Raspbian on your Pi and set up the operating system. Expand the file system and check that it's up to date.

» Configure USB Wi-Fi dongle. Enter `$ sudo nano /etc/network/interfaces` and change the `wlan1` entry to the following:
```
allow-hotplug wlan1
iface wlan1 inet manual
    pre-up iw phy phy1
interface add mon1 type monitor
    pre-up iw dev wlan1 del
    pre-up ifconfig mon1 up
```
Then save changes and reboot.

» Enter either `$ sudo iw dev mon1 set freq 2437` or `$ sudo iwconfig mon1 channel 6`.

2. MONITORING SETUP

» Download, build, and install kismet by inputting the following:
```
$ sudo apt-get install git-core
build-essential
$ sudo apt-get install
libncurses5-dev libpcap-dev
libpcre3-dev
libnl-dev libmicrohttpd10
libmicrohttpd-dev
$ sudo wget
http://kismetwireless.net/code/
kismet-2016-07-R1.tar.xz
$ sudo tar -xvf kismet-2016-
07-R1.tar.xz
$ sudo cd kismet-2016-07-R1
$ sudo ./configure
$ sudo make
$ sudo make suidinstall
$ sudo usermod -a -G kismet pi
$ sudo mkdir -p /usr/local/lib/
kismet/
$ sudo mkdir -p /home/pi/.kismet/
plugins/
$ sudo mkdir -p /usr/lib/kismet/
```
Allow the Pi to reboot.

» Follow the online project page to download the manufacturer list for kismet to identify networked devices.

3. SCAN SETUP

» To install nmap and arp-scan, enter the commands `$ sudo apt-get install nmap` and `$ sudo apt-get install arp-scan`.

» We can update the `mac-vendor.txt`

file to provide more information on the devices detected by executing the following commands:

```
$ cd /usr/share/arp-scan
$ sudo mv mac-vendor.txt mac-
vendor.orig
$ sudo wget http://bit.ly/mac-
vendor
```

» Next, we'll use `arp-scan` to create a log of the devices connected to the network throughout the day. First, we need to install the following packages:

```
$ sudo apt-get install dnsutils
$ sudo apt-get install libdbd-
sqlite3-perl
$ sudo apt-get install libgetopt-
long-descriptive-perl
$ sudo apt-get install
libdatetime-format-iso8601-perl
```

» Then enter `$ sudo wget http:// bit.ly/2v8rRGb` to save the `counter. pl Perl` script to your Raspberry Pi. To ensure that the script is executable, enter `$ sudo chmod uog+x counter.pl`. Now to test functionality, enter `$ sudo ./counter.pl --network home` into the command line.

» Now that we have everything working, in order to have the Pi regularly scan the network we need to add the scan command to crontab. To do so, enter the commands `$ sudo su` and `$ sudo crontab -e` to open up the crontab file, and add the following line to the end of the file:

```
0,30 * * * * /home/pi/counter.pl
--network home
```

» We can make sure the database is updated upon every reboot by adding the following script to the `/etc/rc.local` file:

```
#!/bin/sh -e
#
# rc.local

# su pi -c '/usr/local/bin/
kismet_server -n -c mon1
--daemonize' /home/pi/counter.pl
--network kaleider &

exit 0
```

Save and exit the file.

4. BUILD THE DISPLAY

» Place strips of electrical tape on the back of both Large Digit Driver boards in order to protect the vias (Figure Ⓐ). Align the 10 pins of the boards with the traces at the bottom on the back of the 7-segment displays. Proceed to solder all 12 castellations to the display (Figure Ⓑ), referring to the SparkFun Castellated Vias Soldering Guide (learn.sparkfun.com/ tutorials/how-to-solder---castellated-mounting-holes) if needed.

» Attach the second display to the first with jumper wires (Figure Ⓒ). Connect GND of the OUT on the first display to the GND of the IN on the second display, LAT of the OUT on the first display to the LAT of the IN on the second display, and so on.

» To hook up the displays to the Arduino, connect Arduino pin 6 to CLK, 5 to LAT, 7 to SER, 5V to 5V, Vin to 12V, and GND to GND (Figure Ⓓ).

» Plug the 12V power supply into the Arduino to power the displays. Copy and upload the single digit example code to the board to ensure proper functionality, and repeat for the two digit code.

You can find these steps and the code for using the Large Digit Driver with the 7-segment display in the SparkFun Hookup Guide (learn.sparkfun.com/tutorials/large-digit-driver-hookup-guide).

5. PROGRAM THE ARDUINO

» We can modify the example code so that the Arduino will accept a number through the serial port and then display it. Upload the code from gist.github.com/ aallan/7ae04d27ac19b8ea90e26f8391 f624c2 to the Arduino, open the serial monitor, and enter a number to test this.

6. ARDUINO-PI COMMUNICATION

» Unplug the Arduino from your computer and plug it into the Raspberry Pi. Execute the command `$ ls /dev/tty*` to see the available serial devices. The Arduino will most likely appear as `/dev/ttyUSB0`.

» Enter `$ sudo apt-get install libdevice-serialport-perl` to install the last package needed for the build.

TIME REQUIRED:
» 5–10 Hours

COST:
» $140

MATERIALS
» **7-segment display, 6.5" red (2)** SparkFun #8530 sparkfun.com
» **Large Digit Driver (2)** SparkFun #13279
» **SparkFun RedBoard or Arduino Uno**
» **USB mini-B cable**
» **Power supply, barrel connector** 12V/2A
» **Jumper wires, male to female (6)**
» **Jumper wires, male to male (6)**
» **Raspberry Pi 3**
» **Power supply, micro USB, 5V/2A**
» **USB Wi-Fi dongle** We used Anewish Mini Wireless RT5370

TOOLS
» **Soldering iron**
» **Solder**

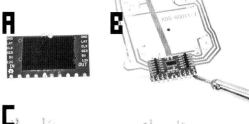

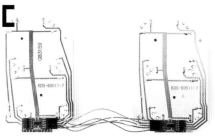

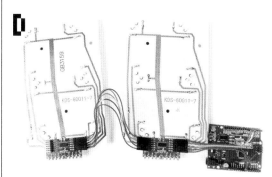

» Replace the previous `counter.pl` script with the updated script found at bit. ly/2uH9JSn.

With that, you should have a configured, working network counter. Mount it in a nice enclosure and display it proudly. Leechers beware — you're being watched! ◗

Written by Brian Lough

TRAVEL LIGHT

Put this light-up NeoPixel map on your wall to see if
the traffic to your destination is green, yellow, or red

TIME REQUIRED:
» 1–2 Days

COST:
» $60–$80

MATERIALS
» **Feather Huzzah ESP8266 and headers** Adafruit #2821 adafruit.com
» **NeoPixel through-hole LEDs (quantity as needed)** We used Adafruit 5mm #1938
» **Capacitor, 220uF**
» **Resistor, 330Ω**
» **Logic Level Shifter** if you add more than five NeoPixels. We used the TXB0104 Bi-Directional Level Shifter, Adafruit #1875 for our Bay Bridge painting, above.
» **Power source, 5V** We used a USB wall adapter.
» **Micro USB to USB cable**
» **Map** printed on paper
» **Computer with internet access**
» **Hook-up wire**
» **Heat-shrink tubing**
» **Breadboard (optional)**

TOOLS
» **Soldering iron**
» **Scissors**
» **Wire strippers**

BRIAN LOUGH is a software developer who has recently gotten into Arduino development after discovering the ESP8266. He posts on his YouTube channel (youtube.com/channel/UCezJOfu7OtqGzd5xrP3q6WA) and Instructables. He lives in Ireland with his fiancée, daughter, and two dogs.

This project displays live traffic conditions between two locations on a physical map, using an Arduino device that gathers data from the Google Maps API and then sets the color of a string of NeoPixels — green for good traffic, yellow for slower traffic, and red for bad traffic conditions, just like online maps. It refreshes every minute to adjust the color of the LEDs as the traffic changes.

The project runs on an Adafruit Feather Huzzah ESP8266 programmed via the Arduino IDE, but the code should run on any ESP8266 board. Its configuration is persistent across resets, as it uses internal flash memory on the ESP8266.

1. BUILD THE CIRCUIT
Place a 220uF capacitor between 3V and GND. With your soldering iron, connect the positive power lead of your first NeoPixel to 3V, and then wire the positive leads of your remaining NeoPixels in parallel. Repeat this same step for GND and the negative power leads.

Place a 330Ω resistor between the output pin of your board and the data-in of your first NeoPixel. Connect the data-out of the NeoPixel to the data-in of the next, and repeat. The data-out of the last NeoPixel should be left disconnected (Figure A).

2. GOOGLE MAPS API KEY
In order to get the traffic data from the Google Maps API you need to use an API key. These are free and very easy to get.

Open developers.google.com/maps/documentation/distance-matrix in your browser. Scroll down to "Quick start steps" and click the "Get a Key" option. Type in the name of your project and click the check box to agree to the terms and conditions. You will then be given your key, which is needed later in the sketch.

I recommend trying out your key using the following URL to make sure it's working correctly (Make sure to change the key at the end!):
https://maps.googleapis.com/maps/api/distancematrix/json?origins=Galway,+Irela

Hep Svadja, Luke Artzt

nd&destinations=Dublin,Ireland&departu
re_time=now&traffic_model=best_guess&k
ey=PutYourNewlyGeneratedKeyHere

If everything appears to be working,
paste your API key into line 89 of the code,
which you'll download from the project's
GitHub page at github.com/witnessmenow/
arduino-traffic-notifier.

3. THE CODE

You now have everything you need to
program your board with the example
sketch. Open the traffic-notifier sketch in
the Arduino IDE. There is a list of libraries at
the top of the sketch — install them if you do
not have them already (all are available on
the Arduino Library Manager).

If you scroll down a bit, you will see a
commented section where you can adjust
the code for your specific use. It is here
that you can set your output pin, number of
NeoPixels, brightness, traffic thresholds,
and refresh rate.

You should now be able to program your
board with the sketch.

4. CONTROLLING THE DEVICE

First, connect the board to Wi-Fi. Power
the board on and find its Wi-Fi network.
Connect your computer to the board's Wi-Fi
and configure it to join your network. Now it
can load traffic condition data.

Find your starting longitude and latitude
on Google Maps — they'll show in the URL.
Copy the coordinates and input them into
line 96 of the code. Repeat this for your
destination (line 97). Upload the code to the
board, then unplug your computer. Supply
power to the board if not already done. Wait
a few seconds as the map downloads the
traffic data and changes the map's lights to
the appropriate color.

5. MOUNT IT

This part is up to you — we used a vintage
map of San Francisco (Figure **B**) that we
downloaded and printed to track traffic over
the Bay Bridge, placing four LEDs along the
bridge's path (Figure **C**). We mounted the
map and circuitry to a foamcore board for
rigidity (Figure **D**).

This can be placed inside a nice picture
frame and hung on the wall (Figure **E**).

> **NOTE:** If you use more than 5 LEDs,
> you may need to add a logic level
> converter to supply adequate power to
> all. Wire it up as shown in Figure **F**.

LIGHT THE WAY

Your traffic map is done! Now you simply
have to plug it in using any standard 5V
power source with a micro-USB plug on
it. The board will connect to the internet
and run the code; a few moments later, the
LEDs will light up to show you the traffic
between the locations you've specified.

The Google Maps API paired with an
ESP8266 has the potential to create some
really interesting projects that represent the
information available on Google maps in a
physical way. It can be used to compare the
travel times of several different routes to or
from work (which I've done at instructables.
com/id/Arduino-Commute-Checker), or
even could compare travel times between
different modes of transport (driving/
walking/public transport etc.), encouraging
people to try alternative ways of getting to
work if driving is significantly slower. ⊘

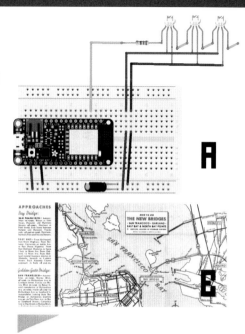

A

B

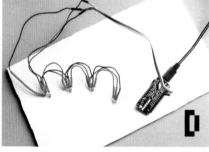

C

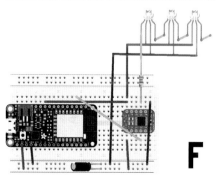

D

E

F

AUTOMATED TOMATO GARDEN

iStock.com/Paul Biryukov, Gordon Williams

TIME REQUIRED:
» 4 Hours

COST:
» $65

MATERIALS
- » **Drip feed kit** including pipe, hose adaptor, T-pieces, and drip feeders. A hose pipe with holes in it works just as well.
- » **Push-fit adaptor**
- » **Puck.js microcontroller**
- » **Battery, 3V coin cell** one comes with the Puck.js
- » **FET, P36NF06L (2)** to turn the solenoid and pump on and off
- » **Diode, 1N4001 (2)** to protect the FETs from the back EMF when the coil is turned off
- » **Battery, 12V lead acid**
- » **Weatherproof box** I'm using a Schneider electricity junction box. They come in various sizes, but mine has a bit of room spare in the bottom.

TOOLS
- » **Drill**
- » **Computer with web access**
- » **Soldering iron and solder**

Use Puck.js to build a smart waterer that fertilizes your plants and more   Written by Gordon Williams

I love tomatoes, especially warm, freshly grown ones. We're lucky enough to have a south-facing garden where tomatoes should thrive given enough water. However, watering is one of those things I've proven myself incapable of doing reliably. After a few false starts with a fish pond pump and a timer, I decided to do the sensible thing and buy a proper plant waterer — which ran out of batteries, got stuck off, stuck on, and finally filled itself up with water. Eventually, being a maker, I decided to do something about it — I made my own waterer with a microcontroller, and it "just worked." Each year it's evolved, and I now have something that's simple, reliable, cheap, and has features you

GORDON WILLIAMS is the author of *Making Things Smart*, which shows how to make simple versions of high-tech items (a TV, camera, printer, etc.) using readily available parts and Espruino micro-controllers.

won't find even in expensive shop-bought waterers, including automatic fertilizing!

ASSEMBLY
The plumbing is easy — drill a few holes in the box (Figure Ⓐ) and put in the solenoid (Figure Ⓑ), then connect it to the hose adaptors (Figure Ⓒ) and watering kit (Figure Ⓓ). Your water source will connect to this. Meanwhile, the peristaltic pump (Figure Ⓔ) will periodically add fertilizer into the water flow; one side goes to a container of liquid tomato feed via a connected hose, and the other plumbs into the watering pipe going to the tomatoes, using a T. If you bought a drip feed kit, they often come with all the pipe and T-pieces you'll need.

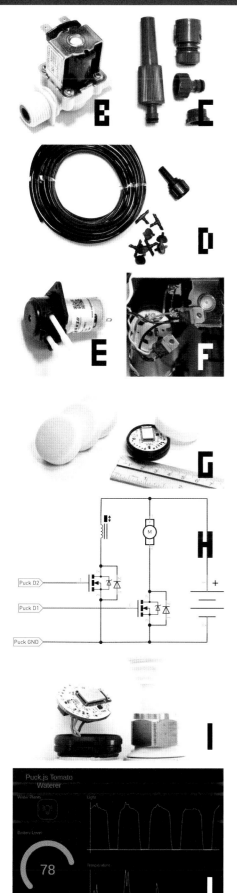

To avoid using circuit boards, I soldered the FETs and diodes (Figure **F**) directly to the solenoid and pump. All you need then is to attach the 12V power and the 3 wires (ground and two control signals connected to D1 and D2) to Puck.js (Figure **G**). Follow the circuit diagram in Figure **H**. I put the Puck.js on top of the case and drilled through for some wires — it's waterproof enough to be fine outside, and it means I can press the button on it to easily start watering when I'm in the area (Figure **I**).

SOFTWARE
Now for the fun bit! Puck.js runs JavaScript, and can be programmed and debugged completely wirelessly using Web Bluetooth.

Follow the instructions at puck-js.com/go to update your Puck's firmware and get the Espruino IDE open — in many cases it'll "just work" as long as you have the Chrome web browser.

Once you have the Web IDE open, click the "connect" icon in the top left and choose your device. On the left side of the IDE you should see a command-prompt where you can instantly interact with the microcontroller itself. You can still write code "normally" on the right-hand side and upload it in one go.

Go into Settings (top right) then Communications, and make sure "Set Current Time" is checked (this will set the correct time whenever you upload code).

Then upload the following code for a simple waterer:

```
E.setTimeZone(-8 /* PST */);
var hadWater = false;

function waterPlants(water, feed) {
  digitalPulse(D1, 1, water*1000);
  if (feed)
    digitalPulse(D2, 1, feed*1000);
}

// Check watering every 10 minutes
function onTick() {
  var now = new Date();
  var h = now.getHours();
  var day = now.getDay();
  if (h==8 || h==19) {
    // feed on mon, wed, fri morning
    var doFeed = (h==8) &&
      (day==1 || day==3 || day==5);
```

```
    if (!hadWater)
      waterPlants(300, doFeed?30:0);
    hadWater = true;
  } else {
    hadWater = false;
  }
}
setInterval(onTick, 10*60000);

// When a button is pressed,
//water for 30 sec
setWatch(function() {
  waterPlants(30,0);
}, BTN, {edge:"rising",
  debounce:50,repeat:true});
```

This will water twice a day at set times, and will also fertilize the tomatoes three times a week (Monday, Wednesday, Friday).

Using the left side of the IDE, you can interact directly with the waterer. `D1.set()` will turn on the water solenoid and `D1.reset()` will turn it off. Typing in `waterPlants(10,5)` will water the plants for 10 seconds and feed them for 5 seconds.

You're now done. Your plant waterer should keep working just fine until the battery runs down in a year or so. However, chances are you'll want it to keep working even if it gets reset, and that's easy — just type `save()` and everything will be saved into flash memory.

> **NOTE:** By default anyone can connect to Puck.js and interact with it. Visit espruino.com/Puck.js+Security for some ideas if you're worried about this!

Check out github.com/gfwilliams/MakeTomatoes for this code as well as a slightly more complicated version that logs temperature and light levels, and even provides a dashboard for your phone (Figure **J**)!

If you attempt this (or something similar) yourself, or have any questions, please get in touch on the Espruino forums: forum.espruino.com. ◗

For code files and more on using the Web Bluetooth interface, go to github.com/gfwilliams/MakeTomatoes.

DIY HOVER PLANT

Build your own magnetic levitating planter
(or whatever floats your boat)

Written and photographed by Jeff Olson

I discovered the Lyfe levitating planter ($229) last year and thought I could make my own for less. I bought a cheap electromagnet on Amazon and posted my first creation on Reddit. Everyone thought it was pretty cool (reddit. com/r/woodworking/ comments/5hku4g) but I wasn't satisfied with the levitation distance and stability of that version, so I set out to find the manufacturer of the same magnet used in the Lyfe planter. It's simply a larger magnet that levitates higher and is more stable in the air. Here's how to build your own:

1. CREATE OPENING FOR POWER SUPPLY
Drill a ½" hole in the back of the cigar box for the power cord to feed through.

2. PLACE BASE AND PLUG
Connect the power cord to the electromagnet base, and close the box. If the cigar box is too deep, you should raise the electromagnet up higher so it is as close as possible to the lid. I cut out a piece of MDF and glued it to the bottom of the cigar box to close the distance.

3. REMOVE CAN TOP
Cut the top off of the old steel beer can using a kitchen can opener.

4. ATTACH MAGNET DISC
Stick the levitation disc magnet to the bottom of the can. With a steel can, there's no need for glue since the magnet will hold tight. If you use another container, you'll need to glue the levitation magnet to the bottom.

5. ADD YOUR GREENERY
Place your air plant inside your planter.

USE IT
It can be a little tricky to find the levitation sweet spot, but with practice it becomes easy. Make sure the base is plugged in, and keep metal objects away so as not to interfere with the magnet. Hold the magnetic disc about 15cm (6") above the base. Lower the disc with both hands directly over the center of the base, keeping it level until you feel the upward magnetic force supporting the weight of the disc. Gently let go, keeping it centered and level.

If it falls, simply lift the disc up and try again. Expect that it will take several attempts and may require some practice to master.

Your hover planter will gently spin for hours, allowing the air plant to receive 360° of sunlight.

The magnet I used is completely silent; there's absolutely no hum or electronic noise. More importantly, it just looks cool! ◆

CAUTION: Do not levitate the disc without placing some sort of padding over the electromagnet base (like your cigar box lid). The exposed magnets can be damaged if they slam directly into each other.

Watch the DIY Hover Plant in action, and share your build at makezine. com/go/diy-hover-plant.

JEFF OLSON lives in Denver, Colorado, where his hobbies are CNC projects and Red Rocks concerts. He shares his new "levitation" projects at levdisplay.com.

TIME REQUIRED:
» 1–2 Hours

COST:
» $140–$160

MATERIALS
» **Cigar box**

» **Beer can or other small container** for a planter. I used an old steel beer can, so the levitation magnet sticks to it without any glue. If you use a non-steel container, you'll just need to add some glue.

» **Magnetic levitation device, 350g (13oz) capacity** includes electromagnet base and separate disc magnet for levitation, as well as power supply. I tracked down the same hardware that's in the Lyfe planter; you can buy it from me on Amazon, #B06XSN5CX4, or at my store levdisplay. com. Alternative electromagnets will work for this project, but typically have a shorter levitation distance and less stability (meaning the planter can be thrown off balance more easily, causing it to wobble or fall).

» **Small plant** It's possible to use small succulents (which require soil) but I highly recommend using air plants (*Tillandsia*) since they're much lighter and require no soil and little water.

TOOLS
» **Drill with ½" bit**

» **Can opener**

» **Glue (optional)**

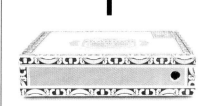

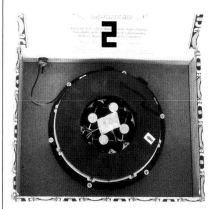

Heat Vision

My path to building a low-cost, DIY thermal imager kit

Written by Max Ritter

MAX RITTER is an engineer who loves everything about technology and is passionate about doing new things with unconventional ideas.

THE BASIC IDEA OF MY LOW-COST THERMAL IMAGER STARTED WITH MY PHYSICS CLASS in 2010. Our teacher bought a single-point infrared thermometer, also called a pyrometer, and asked if anyone wanted to use it for a science competition later that year. A friend and I came up with the idea of creating a thermal image scanner using servomotors to move the infrared sensor over a large area.

Our first prototype was nothing more than a proof of concept; we used Lego Mindstorms together with a data interface for the sensor on the PC and an automated mouse/keyboard script for Adobe Photoshop to create low-resolution thermal images. I improved our design for the next year's competition (Figure Ⓐ), which I called the "Cheap-Thermocam V1." It consisted of an Arduino microcontroller, two servomotors, and computer software written in Java; the total material cost was only about $100.

FIRST KITS

In 2011, I won a special prize and published the software and hardware concepts to the internet. The feedback was positive — many people built their own adaptation of the device (Figure Ⓑ, following page), and others expressed interest in buying one. This led me to the second version of the device (Figure Ⓒ, following page) in mid-2013, which I sold online to people from all over the world. I was 20 at that time, so this was a huge step for me. The "Cheap-Thermocam V2" featured a small LCD display, which could be controlled over a rotary encoder and had the option to save data to an SD card.

A year later, I finished my work on the "Cheap-Thermocam V3 (Figure Ⓓ, following page)," which integrated a large touch screen, a slim design, and a much faster microcontroller. Versions one to three used the original principle of a movable single-point IR sensor that scans an area.

**Time Required:
2–4 Hours
Cost:
$500–$550**

MATERIALS

» **DIY Thermocam Kit** groupgets.com The self-assembly kit includes all required parts except the Lepton sensor and module. After you click on "Join this buy," you can add the Lepton module and the Lepton 3.0 or Lepton 2.5 sensor to your order.

OR, GATHER THE FOLLOWING PIECES:

» **FLIR Lepton (shuttered) long-wave-infrared array sensor** v2.0, v2.5, or v3.0
» **FLIR Lepton breakout board interface**
» **Spot infrared sensor** Melexis MLX90614-BCF, used for absolute temperature measurement (not required for Lepton 2.5)
» **Teensy 3.6 microcontroller board** Arduino compatible, pjrc.com
» **MicroSD card, Class 4, 8GB**
» **ArduCam Mini V2 camera module, 2MP**
» **Display module, 3.2" TFT LCD** Configuration: 5V, pin header 4-wire SPI, resistive touch, no font chip.
» **Battery, 3.7V lithium polymer (LiPo) with JST-PH connector** max. dimensions 60mm×55mm×6.5mm high
» **Printed circuit board** 89.4mm×68.4mm, 1.6mm thick, 2 layers. Get the Gerber files at github.com/maxritter/DIY-Thermocam/tree/master/PCB.
» **Enclosure, laser-cut from 3mm black acrylic plastic** Get the design files at github.com/maxritter/DIY-Thermocam/tree/master/Enclosure.
» **Battery charging module, TP4057** includes charging LED
» **Voltage booster, 5V, U3V12F5** aka step-up voltage regulator, Pololu #2115
» **Power switch** E-Switch #R6ABLKBLKFF
» **Pushbutton** RAFI #1.10107.0110104
» **USB power switch** E-Switch #EG1201A
» **Battery connector** JST #S2B-PH-K-S, connects the LiPo to the PCB
» **SD card slot adapter** Wurth #693063020911
» **MicroSD adapter**
» **Display connector, 40-pin, 2.54mm female header**
» **Lepton board connector, 8-pin, 2.54mm female header**
» **Pin header strip, 40-pin, 2.54mm male header**

» **Coin cell battery holder** Keystone #3001
» **Battery, CR1220 coin cell** for the real-time clock
» **Resistors, 4.7kΩ ¼W 1% (4)**
» **Resistors, 10kΩ ¼W 1% (2)**
» **Tape, double-sided**
» **MicroUSB cable, angled**
» **Tripod, mini**
» **Tripod socket, ¼-20**
» **Screws, M2×10mm (6)**
» **Screws, M2×8mm (5)**
» **Screws, M2.5×6mm, black (8)**
» **Spacers, M2×3mm(6)**
» **Spacers, M2×3.5mm (2)**
» **Standoffs, M2.5×12mm (4)**
» **Standoffs, M2.5×11mm (4)**
» **Standoffs, M2.5×5mm (4)**
» **Nuts, M2 (5)**
» **Nuts, M2, plastic (6)**
» **Washers, M2 (3)**

TOOLS
» **Soldering iron and solder**
» **Pliers**
» **Wire snips**
» **Screwdriver**
» **Multimeter**

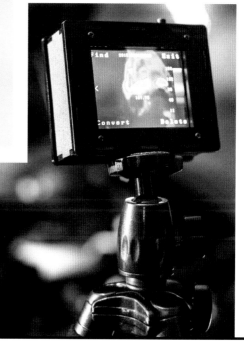

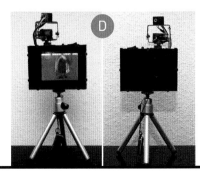

Terence Musho, Max Ritter, George Rhoten, Courtesy of Max Ritter

While this had the advantage of being very cheap, it also took a couple of minutes to create one full thermal image. For many applications including moving objects, this approach was not suitable, so I was looking for an alternative.

A NEW SENSOR

In 2014, FLIR released their Lepton sensor (Figure **E**), which was the first low-cost thermal array sensor on the market. I included it in the next version of my device, which I called the "DIY-Thermocam V1." It was a big improvement from the scanning-principle to high-resolution, real-time thermal images and the starting point towards a serious alternative to the solutions on the market from big companies like FLIR or FLUKE.

Version 2 of the DIY-Thermocam provides many enhancements, including support for the Lepton 3.0 sensor with four times increased resolution, a more powerful microprocessor, a better and faster visual camera, and removable storage. A new video output module offers the option of streaming a video output signal of the data from the Thermocam.

A big advantage of the Thermocam is the open-source software and hardware, so you can modify it to your own needs or use it as a starting point for your own developments in the field of thermal imaging. The on-device firmware can be controlled with an easy-to-use touch menu and offers many features, like different color schemes, analysis, or saving methods. For analysis of the thermal images and videos on the PC, the raw files are fully compatible with a powerful application called ThermoVision, developed by a German programmer from Berlin. There's also a Python application that offers real-time streaming of the thermal data on the computer and has some analysis features.

LESSONS LEARNED

There were so many challenges along the way. In most cases I was able to solve them, in others I decided to try more unconventional solutions. For example when designing the case, I first wanted to use a 3D printer to produce it. However, it turned out that even the more expensive printers required a lot of post-printing

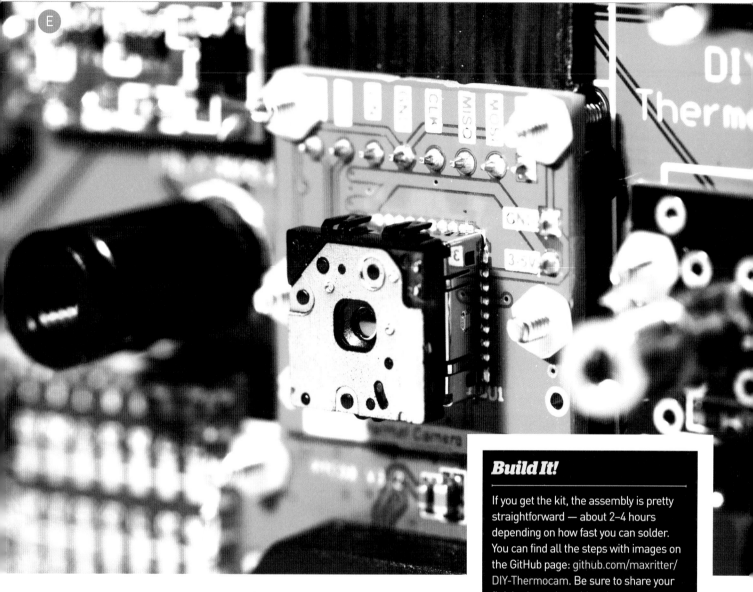

Build It!

If you get the kit, the assembly is pretty straightforward — about 2–4 hours depending on how fast you can solder. You can find all the steps with images on the GitHub page: github.com/maxritter/DIY-Thermocam. Be sure to share your finished creation with us!

refinement by hand. This took too long to be productive. Injection molding was not a real alternative either, as creating a mold was excessively expensive for making a small number of units. In the end, I decided to laser cut an enclosure, a relatively cheap technique that requires no post processing. I found a company in Germany who cut the parts for me out of black acrylic.

One thing I learned was not to set the time schedule for the release too tight, because there are many unexpected problems that occur. A good rule of thumb I found for myself is to make a schedule and then double that amount. If you are done earlier, that often isn't a problem, because you always find something to improve on.

I learned all the things I needed for this from sources on the internet and books.

There was almost nobody that helped me with it. This was a steep path, but I learned so much from it! The project contained many steps like making the prototype, choosing the components and materials, making the device easy-to-assemble, quality checkups, worldwide delivery, marketing, and customer contact, just to name a few. Many of the things I did were trial and error, and I really made plenty of mistakes during those years. However, all of them contributed to the treasure of experiences I have today, so it was worth making all of them. I would even say that I learned more working on the Thermocam project than from my five years of studies. In my opinion, you make the most progress — both personally and technically — when you actually work on challenges by yourself. ◑

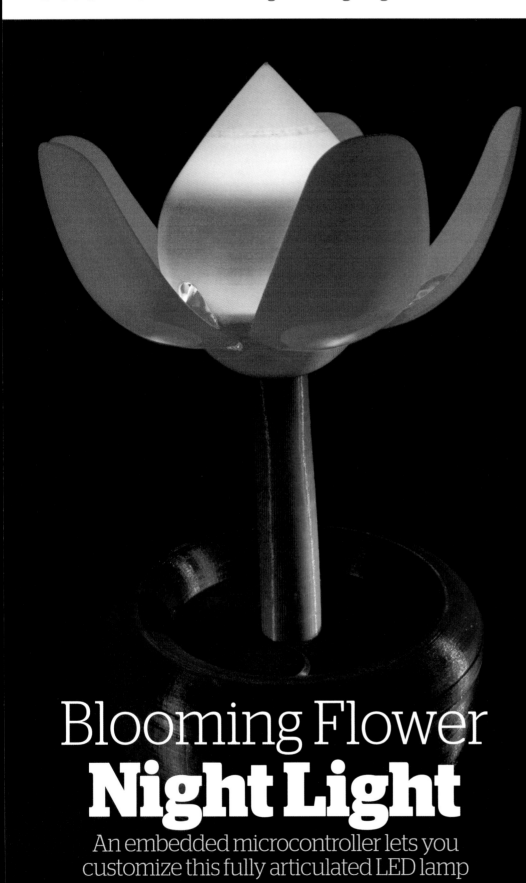

JASON SUTER is an engineer by trade and tinkerer by heart. He has been making as long as he can remember, with a particular passion for things that move. 3D printing has opened a whole world of possibilities for him, especially to tackle custom R/C vehicles that he never had the means to create before.

Time Required:
1 Week
Cost:
$50–$70

MATERIALS
- » **Arduino Nano microcontroller**
- » **Servo, R/C, 9g** or other small servo
- » **NeoPixel Rings, 16-LED (2)** Adafruit #1463, adafruit.com
- » **Cellphone charger**
- » **USB cable**
- » **Rotary encoder with push button, 12mm shaft** without detents may be preferable
- » **Nuts, M3 (5)**
- » **Screws, M3: 6mm (5), 10mm (20), and 8mm countersunk (9)**
- » **Pins, 3mm×15mm (5)** cut from 3mm smooth or threaded rod, or screws with heads cut off
- » **Cable, stainless steel, 1mm– 2mm diameter (~20cm)**
- » **Pushrod connector** for standard servo, such as Du-Bro or Sullivan

3D-PRINTED PARTS
- » **STL files** Get them at myminifactory.com/object/37752.
- » **Filament** PETG is recommended

SOFTWARE
- » **Arduino IDE** free from arduino. cc/downloads
- » **Adafruit NeoPixel and TiCoServo libraries** free from github.com/adafruit/Adafruit_ NeoPixel and learn.adafruit.com/ neopixels-and-servos/the- ticoservo-library

TOOLS
- » **3D printer with at least 6"×6" bed size**
- » **Super glue**
- » **Hacksaw**
- » **Drill bit, 3mm–3.5mm** to clean out holes
- » **Double-sided tape/servo tape**
- » **Screwdriver**
- » **Hobby knife or scalpel** for cleaning up parts
- » **Soldering Iron and solder**
- » **Zip ties (optional)**

Blooming Flower
Night Light

An embedded microcontroller lets you customize this fully articulated LED lamp

Written by Jason Suter

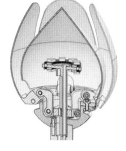

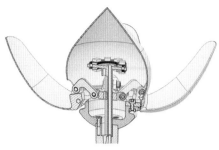

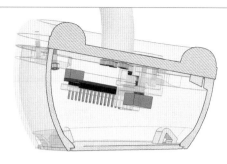

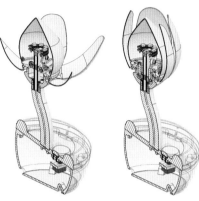

3D-printed parts

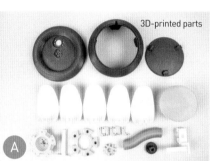

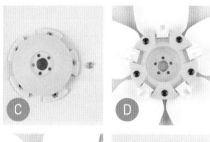

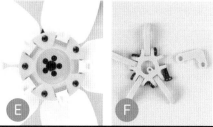

I WAS ASSIGNED THE SIMPLE TASK OF MAKING A NIGHT LIGHT for my soon-to-be-born baby girl's room and, as is my norm, I got a bit carried away, ending up with a fully articulated, Wi-Fi-enabled, 3D-printed night light. Unfortunately the Wi-Fi idea was over the top, and it never actually got used. There was also far too much hot glue under the hood for my liking, so this revision is a cleaner, simpler version operated with a push button, making use of easily sourced components from Arduino and Adafruit.

The heart of this blooming flower is a cable-actuated mechanism, which slides up and down a shaft in the center of the bloom, pulling and pushing linkages that lever the petals in and out. Full-color illumination is provided by two NeoPixel LED rings mounted back to back.

PRINTING THE PARTS

Get the STL files at myminifactory.com/object/37752 and 3D-print them (Figure A). Although almost any material should work, PETG is the most suitable due to is flexibility and strong interlayer adhesion, both of which benefit thin features (like the petals), which can take some abuse. For more tips about printing the parts, visit this project online at makezine.com/go/blooming-flower-night-light.

MECHANICAL ASSEMBLY

The mechanism and electronics all attach to the top of the pot, allowing easy access to all components until the system is complete, at which point the bottom of the pot can be installed. You can find an assembly video of the first iteration on the project page online.

NOTE: It's imperative that all pivoting parts can move with as little resistance as possible. If necessary, use a 3.5mm drill bit to clear and enlarge the holes.

1. TEST-FIT THE PETALS
Ensure that the petals turn smoothly on the 3mm pins (Figure B) — clean out the holes if necessary. If you can't find smooth 3mm rod, then 15mm–22mm lengths of threaded rod or screws cut with a hacksaw will work just as well. Remove the petals.

2. INSERT THE M3 NUTS
Place an M3 nut into each of the five hex-shaped recesses in the "receptacle lower" part (Figure C). If they're too tight due to printer tolerances, carefully pressing them into place with a hot soldering iron works very well.

3. AFFIX THE PETALS AND STEM
Hang the petals into the "receptacle lower" part by the 3mm×15mm pins. Place the "receptacle upper" part over the pins. Fasten the five M3×6mm screws through both parts into the M3 nuts, locking the pins in place (Figure D). Ensure that the petals still move without resistance, then attach the "stem" to the receptacle with the five M3×8mm countersunk screws (Figure E).

4. ADD THE LINKAGE ASSEMBLY
Attach the long side of the five L-shaped "linkage" parts to the "linkage star" with M3×10mm screws (Figure F). The linkages should pivot freely on the screws, which are tapped directly into the linkage star.

Screw the five linkages into the petals with M3×10mm screws, once again checking that the linkages can pivot freely on the screws (Figure G on the following page). Make sure that the hole for the cable in the linkage star and in the stem line up.

5. ATTACH STEM TO POT
Affix the "stem" to the "pot upper" (top of the flowerpot) with five M3×10mm screws (Figure H on the following page). Insert the "axis" part through the linkage star and into the receptacle, but do not yet glue in place.

6. ADD CABLE AND TEST MECHANISM

Thread the stainless steel cable through the linkage star and the axis, and down the stem. Use a drop of super glue to lock the cable into the star (Figure **I**). Leave extra cable at the bottom of the pot (Figure **J**), but trim the top of the cable.

It's important to confirm that the linkage moves smoothly and without resistance by gently pushing and pulling the cable at the bottom of the stem (depending how loose it fits, you may need to hold the axis, shown in Figure **K**, in place with your finger).

7. ATTACH THE AXIS

Remove the axis, place it through the lower NeoPixel ring, and reinsert it, using a drop of super glue to hold it in place in the receptacle. Note that the lower NeoPixel ring must be installed first (facedown), since it will not fit over the top of the axis.

ELECTRONICS ASSEMBLY

This project uses an Arduino Nano, but it would be easily adaptable to almost any microcontroller, especially one that's compatible with Adafruit's Arduino libraries. To program your Nano, find the Arduino IDE at arduino.cc/downloads.

The wiring (Figure **L**) is simple. Three separate subsystems — servo, LED rings, and rotary switch — each communicate with the Arduino and can be tested individually.

1. INSTALL THE LIBRARIES

Use the Arduino Library Manager to install:
» Adafruit NeoPixel library
» TiCoServo libraries

2. ATTACH THE MICROCONTROLLER

The Arduino Nano is simply zip tied or double-sided taped to the "electronics bracket." Its orientation is determined by whether you prefer to use the board "pins up" and intend to use jumpers (Figure **M**) or "pins down" and intend to solder directly to the PCB. It is easier to see the markings on the board in "pins down" orientation.

3. INSTALL THE ROTARY ENCODER

Use the supplied nut to attach the rotary encoder to the "electronics bracket" (Figure **N**).

4. ATTACH THE BRACKET

Affix the bracket containing the encoder and microcontroller to the flowerpot top with 4 countersunk M3×8mm screws (Figure **O**).

5. AFFIX THE SERVO

Attach the pushrod connector to the servo horn and place the horn on the servo — but don't screw it down yet. Use strong double-sided tape to attach the servo to the "pot upper" such that the pushrod connector lines up with the cable hole in the stem. The servo horn can be screwed on after the servo's range of motion has been tested in Step 7.

6. WIRE THE SERVO

Since the Adafruit NeoPixel libraries conflict with the standard Arduino servo libraries, it is imperative that a hardware PWM-capable pin is used for the servo so that the "TiCoServo" library can be used. Connections are:
» 5V on Arduino to 5V on servo
» GND on Arduino to GND on servo
» D10 on Arduino to Signal on servo

7. TEST RANGE OF MOTION

Use the sketch *servoLimitTest* at github.com/ossum/bloomingossumlamp to determine the servo positions that correspond to open and closed. First center the servo without attaching the servo horn, to prevent damaging it by mistake. Instructions on using the sketch are contained in the sketch itself as well.
» Make sure that the **servoPin** variable refers to the pin to which your servo is connected
» Upload the sketch to the Nano, then open the Arduino serial monitor, and
  » enter the characters **q** or **e** to test the servo's outer limits
  » enter **w** to move to the mid point
  » enter **o** or **p** to step in either direction (up to the limit)
» The servo's current position is reported in the serial monitor, so you can determine what limits you would like to use in your final code.

8. WIRE THE NEOPIXEL RINGS

The NeoPixels are actually strands of individually addressable WS2218 RGB LEDs, which means they can be chained by connecting the Data Out (DO) of one ring to the Data In (DI) of the next. Thread three strands of wire through the stem (Figure **P**)

and connect them as follows:
- » 5V from Arduino to Vin on both rings
- » GND from Arduino to GND on both rings
- » D4 on Arduino to Data In (DI) on first ring
- » Data Out (DO) on first ring to DI on second ring

9. TEST THE LEDS

It is wise to test each element of the system as you go along. Use Adafruit's example sketch *strandtest* (github.com/adafruit/Adafruit_NeoPixel/tree/master/examples/strandtest) to confirm that your LED rings are working (Figure **Q**). Make sure to set the pin number in the sketch correctly (D4 in this case). Now you can mount both rings to the axis using bits of double-sided tape.

10. WIRE THE ROTARY ENCODER

Since rotary encoders can have different pinouts, you'll need to read the datasheet or markings on yours to determine which pins are which. The code uses hardware interrupts to handle the encoder's movement, so it is important to use pins on the Arduino that support hardware interrupts:

- » D2 from Arduino to pin A on encoder
- » D3 from Arduino to pin B on encoder
- » GND from Arduino to GND on encoder
- » D7 on Arduino to SW on encoder
- » GND from Arduino to GND on encoder switch

Once the encoder is wired up (Figure **R**) you can test it using the final *FlowerLamp* sketch (github.com/ossum/bloomingossumlamp), making sure that the pins are allocated as you chose them. Watch the Arduino's serial monitor to see that turning the encoder and pressing the switch cause output messages.

11. ATTACH THE BASE AND BULB

Once it's all working, attach the "pot lower" with 10mm screws and snap the "base lid" in place. Finally, snap the clips on the diffuser "ball" to the receptacle to hold it in place (Figure **S**).

INVENTIVE ILLUMINATION

Since the lamp receives all of its power from the 5V USB input to the Arduino, it can be powered by an old cellphone charger. In use the lamp is designed to be as accessible as possible, with a simple press of the button turning it on or off and a twist of the knob to change the color.

There is endless room for customization too, making it a great test bed for learning a bit of coding. Perhaps you'd like the lamp to turn off by itself after 30 minutes, or the colors to slowly change. Or maybe you want to add a Wi-Fi module and scrape weather pages for sunset times, turning the lamp on as the sun goes down. ◐

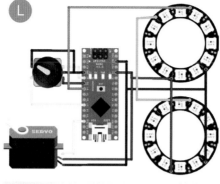

L

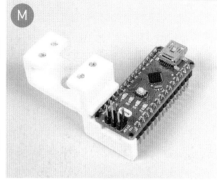

M

N

O

P

Q

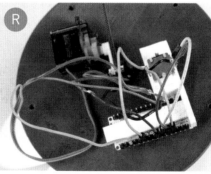

R

S

Jason Suter / created with Fritzing

Annoy-O-Bug

Prank your friends by hiding this cheap and easy chirping, blinking throwie

Written by Alex Wulff

ALEX WULFF is an 18-year-old app developer and maker from upstate New York. He loves embedded systems and applying hardware to address problems in his community. See alexwulff.com for some of his builds.

Time Required:
1–2 Hours
Cost:
$5–$10

MATERIALS

» **PCB, custom fabricated** Buy it from OSH Park, or download the Gerber files, at oshpark.com/projects/XoCU9Yxf.

» **Piezo buzzer, 12mm diameter**

» **LED, through-hole**

» **Resistor, 330Ω, ⅛W** Important: The PCB fits a ⅛W resistor only.

» **Coin cell lithium battery, 3V, CR2032**

» **Battery holder, CR2032 size, clip-style** such as Amazon #B00GYW39KG, amazon.com

» **Atmel ATtiny85 microcontroller IC chip, in DIP-8 package**

» **DIP-8 socket** Choose one that has a hole in the middle.

» **Slide switch, 3-pin SPDT, breadboard-friendly**

ONLY NEEDED FOR PROGRAMMING ATTINY:

» **Arduino Uno**

» **Capacitor, 10µF**

» **Jumper wires**

TOOLS

» **Soldering iron with a fine- to medium-sized tip**

» **Flush cutters or diagonal cutters** to clip component leads

» **Solder**

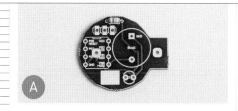

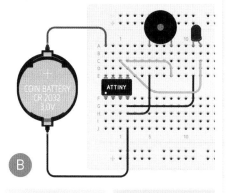

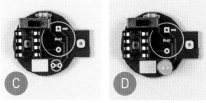

THIS IS A TINY CIRCUIT THAT ANYONE CAN BUILD TO ANNOY FRIENDS AND FAMILY. It features a piezo buzzer to provide unwanted auditory stimulation, and an LED to blink in the wee hours of the night. The circuit board is small and simple enough that anyone can assemble it in a matter of minutes, yet diverse enough that you can have fun creating different programs for weeks. Improve your soldering, programming, and inventing skills with this easy project!

1. ORDER THE CIRCUIT BOARD
You can purchase the printed circuit board (Figure Ⓐ) from OSH Park at oshpark.com/projects/XoCU9Yxf. You by no means need to use an ATtiny or my own PCB; this circuit would take a grand total of about 20 minutes to assemble on any breadboard. If you want the circuit to be tiny, however, I would go with the PCB option. If breadboarding it or perf boarding it does interest you, the breadboarded version is shown in Figure Ⓑ. You could swap out the ATtiny85 for any microcontroller.

2. SOLDER THE NON-POLARIZED COMPONENTS
The slide switch, resistor, and DIP socket in this build are all non-polarized. This means that the orientation in which you solder them doesn't matter. Solder these onto the PCB (Figure Ⓒ) and make sure the connection is solid, then clip the excess leads off the back of the PCB.

3. SOLDER THE POLARIZED COMPONENTS
The LED (Figure Ⓓ) and buzzer (Figure Ⓔ) are both polarized, so the orientation of these components does matter. The shorter, negative (–) lead of the LED goes through the hole opposite the white rectangle on the PCB, and the longer, positive (+) lead through the closer hole. The shorter lead of the buzzer (the side marked negative) goes through the square solder pad, and the positive lead through the round pad. Clip off the excess leads once again.

4. SOLDER THE BATTERY HOLDER
The only tricky part of this project is getting the battery connector soldered. Make sure you solder all the components on the front first. You can then solder the ground pin of the battery connector (Figure Ⓕ) to its pad through the hole in the center of the DIP socket. The positive lead is easy by comparison. Now place the battery.

I have also included a white silk-screen rectangle on the PCB in case you want to write your prankee a small message. If you would like to customize the PCB, just duplicate my circuits.io design at circuits.io/circuits/2677013-annoying-circuit first.

5. EXPLORE THE SOFTWARE
I have created a variety of different example sketches that you can use with this circuit. These can be found at github.com/AlexFWulff/Annoy-O-Bug. Don't just use my provided sketches, though! Try and come up with some of your own. For example, you could improve upon my initial design by using the LED as a light sensor to only chirp at night.

6. PROGRAM THE ATTINY
Before placing the ATtiny85 into the DIP socket, plug it into a breadboard first so you can program it (Figure Ⓖ). There are a wealth of tutorials online that show you how this can be accomplished with an Arduino Uno, but I like this one in particular: create.arduino.cc/projecthub/arjun/programming-attiny85-with-arduino-uno-afb829.

7. PUT THE ATTINY85 INTO YOUR PCB
The orientation of the ATtiny85 in the DIP socket is extremely important. The dot in the top left corner of the ATtiny85 needs to be on the side of the socket closest to the slide switch and not the LED (Figure Ⓗ).

GET PRANKIN'
You now have a fully functional annoying machine! Depending upon where you source your parts these babies can cost you less than $5; it's relatively inexpensive to make many of them.

Planting this device is half the fun. They're small enough to be placed in potted plants, small boxes, pillows, inside lamps, on desks, and anywhere else you can imagine! Add a magnet and try throwing it someplace metallic and out of reach. If you use a watchdog timer to put the ATtiny to sleep, your circuit can run for over a year on a coin cell battery. ◗

Three-Pendulum Harmonograph

Build a swinging art table for uniquely hypnotic drawings

Written and photographed by Karl Sims

KARL SIMS is a digital media artist, and visual effects software developer. He founded GenArts and is a graduate of MIT.

**Time Required:
A Weekend
Cost:
$200-$250**

MATERIALS

» **Plywood, ¾"×3'×3'** for tabletop
» **Lumber, 1½"×1½"×40" (4)** aka 2×2 lumber, about 14' total, for table legs
» **Wood boards, ½" or ¾" thick, 8"×12" (4)** about 4' total, for leg braces
» **Wooden dowels, ¾"×4' (4)** for pendulum and pen lifter
» **Oak, ¾"×1½"×30"** to cut for pendulum supports, jam cleat, etc.
» **Hardboard or plywood, ⅛"×11"×11"** to hold paper
» **Metal pipe nipples, ¾"×5" (3)**
» **Metal pipe bushings, ¾" to 1" (3)**
» **Steel clamps, 1" (3)**
» **Metal plates, 1¼"×4" (4)** You can cut two 1¼"×8" plates cut in half.
» **Large metal washer, 2½" OD, 1" ID** for gimbal
» **Screw eye** for pen lifter
» **Screws, #10, various: 1", 1¼", 1½", 1¾", 2", 3"**
» **A few thin nails**
» **Weights, 2½lb with 1" hole (8 to 12)**
» **Balsa sticks, ½"×¼"×30" (2)** and maybe a spare or two
» **Various pens** I like the Silver Uni-Ball GEL Impact, Staedtler Triplus Rollerball, Pigma Graphic 1, and Sakura IDenti-Pen. Wide pens or thin markers work best.
» **String**
» **Rubber bands**
» **Paper, 8½"×11" or 9"×12"** black and white

TOOLS

» **Drill and bits, including ¾" and ⅛"**
» **Forstner bit or hole saw, 3"** or you could use a jigsaw
» **Saw**
» **Hammer**
» **Tape measure**
» **File**
» **Sandpaper**
» **Tape**
» **Glue**
» **Drill press (optional)**

HARMONOGRAPHS ARE MECHANICAL DEVICES THAT DRAW PICTURES by swinging pendulums, believed to be invented in 1844 by Scottish mathematician Hugh Blackburn. In my 3-pendulum rotary harmonograph, two lateral pendulums swing back and forth at right angles to each other (one side to side, another front to back) with arms connecting to a pen. A third "rotary" pendulum moves the paper by swinging on any axis or in circular motions.

This harmonograph gives a wide variety of pleasant results, and is fairly easy to build. It's a great project to do with kids and can result in endless experiments creating new types of geometric designs. Here's how you can create your own.

THE HARMONOGRAPH CONSISTS OF:
THE TABLE

The legs are about 37" long (go shorter if you plan to fit this through a doorway) and splay out slightly to allow the rotary pendulum to swing without hitting a leg. The legs are attached to the tabletop and supported with braces (Figure A).

PENDULUM HOLES

Three 3"-diameter holes are drilled through the tabletop for the pendulums. The hole for the rotary pendulum should be centered in a corner about 8" from each edge, just clear of the leg brace underneath. The other two holes should be aligned with the first, about 8" from the edge shared with the rotary pendulum hole, and 3" from the opposite edge (Figure B). You'll need a special, large circular drill bit (Forstner bit) for this, or a hole saw. Or drill a small hole, then cut a wider opening with a jigsaw.

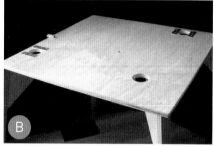

SUPPORT PLATES

Metal plates are mounted on each side of the two lateral pendulum holes. A small indentation is drilled in the center of each plate (Figure C), but unless you have a good drill press, it may be easier to position these indentations on the table after you create the fulcrum blocks with protruding screws, in order to accurately align them.

PENDULUMS AND FULCRUMS

The pendulum shafts rest on hardwood oak block fulcrums, allowing them to rock on the table with minimal friction. For the two lateral pendulums, use 5"-long blocks, and for the rotary pendulum, use a 2¼"-long block. Drill a ¾" hole through the center of each block, and place a 1¼" #10 screw through each end. If you haven't already created the indentions in the metal plates, start them now with a small drill bit (such as ⅛") and then continue with a larger bit (such as ¼"). Be careful not to drill all the way through. Rest the screw tips in the indentations.

To create the pendulums, insert a wooden dowel through the ¾" hole in each fulcrum block (Figure D) such that the screw tips are 12" from the top of the dowel and facing downward. The pendulum should hang down 36" below the top surface of the table with about a 1" clearance from the floor. Glue and/or screw the dowels in place.

GIMBAL

A gimbal allows the rotary pendulum to swing in any direction. Mount blocks (you may want to sand down the edges to allow for clearance) to the underside of the table with screws protruding upward at an angle from either side (Figure E).

Drill a matching pair of indentations into the bottom of the washer, then drill a second pair on top, offset 90° from the bottom pair (Figure F). The washer rests on the screws, and the pendulum rests on the washer.

WEIGHTS

Slide your weights onto metal pipe nipples secured with a bushing screwed onto the lower end, and slide those onto the pendulum dowels. A steel clamp underneath the weights keeps them from sliding off, and allows easy height adjustment to give different swinging frequencies (Figure G, following page).

PAPER PLATFORM

Cut about 1" off the top of the rotary pendulum dowel, so it is slightly lower than the other two. Then mount the 11" square board to the top of this pendulum, using a small oak block glued to it for support, with a ¾" hole for the dowel. Wrap some tape around the top of the dowel to get a tight fit, or just glue it on (Figure H, following page).

Use rubber bands to secure the paper (Figure I, following page), or small clips.

A

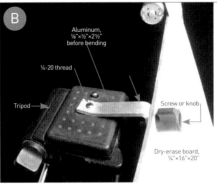

B

Aluminum,
⅛"×½"×2½"
before bending

¼-20 thread

Tripod

Screw or knob

Dry-erase board,
¼"×16"×20"

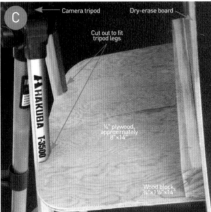

C

Camera tripod

Dry-erase board

Cut out to fit
tripod legs

¼" plywood,
approximately
8"×14"

Wood block,
¾"×1½"×14"

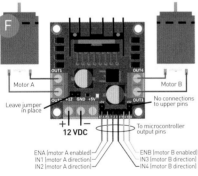

D

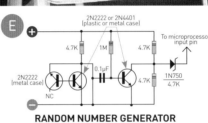

E

2N2222 or 2N4401
(plastic or metal case)

4.7K   1M   4.7K

To microprocessor
input pin

0.1µF

2N2222
(metal case)

4.7K

1N750
4.7K

NC

RANDOM NUMBER GENERATOR

F

Motor A

Motor B

OUT1   OUT4

OUT2 +12 GND +5V   OUT3

Leave jumper
in place

No connections
to upper pins

+   −

12 VDC

To microcontroller
output pins

ENA (motor A enabled)
IN1 (motor A direction)
IN2 (motor A direction)

ENB (motor B enabled)
IN3 (motor B direction)
IN4 (motor B direction)

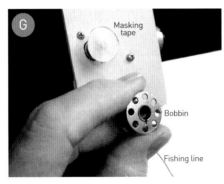

G

Masking
tape

Bobbin

Fishing line

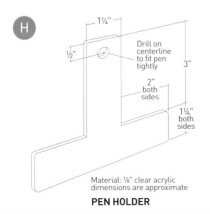

H

1¼"

½"

Drill on
centerline
to fit pen
tightly

3"

2"
both
sides

1¼"
both
sides

Material: ⅛" clear acrylic
dimensions are approximate

PEN HOLDER

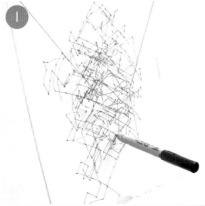

I

Larry Cotton, Juliann Brown, Larry Cotton

shop (Figures B and C). Use a vise and hammer to bend a piece of aluminum to fit between the tripod's camera-mounting screw and your screw or knob at the top of the dry-erase board. Drill and tap the holes to match those threads. Be sure to lean the board back so the pen can draw properly; more on that later. There's a simpler tripod easel online at instructables.com/id/Camera-Tripod-to-Art-Easel that's not as big or sturdy.

Platt and Logue used two 9V batteries to power their RNG. With no 9V batteries on hand, I found a 115VAC supply that provides a steady 12VDC at 1A, which is enough for this project. The motors draw about half an amp total when both are pulling the pen over the drawing surface. The L298N also runs on 12VDC.

Once your parts arrive, build the circuits (Figures D, E, and F). I built mine on top of a BASIC Stamp Homework Board, but you can use any breadboard or proto board. Note that the first transistor (it generates the random noise) has only two leads connected; clip off the third lead. You should read Platt and Logue's detailed build description, but here's an important tip: "When installing the Zener diode, remember that its cathode stripe should point away from the negative bus, unlike a regular diode. The Zener shunts voltages above 4.7VDC to ground."

NOTE: You can safely ignore the right-hand breadboard in Figure D; it comes standard with the HomeWork board, but I played it safe and put my RNG components on the separate breadboard at left, so I didn't accidentally blow my HW board. The (unconnected) pull-down resistors are for insurance; sometimes you gotta guarantee a 0 when you want a 0. The LED is just for testing.

Mount the motors to the left and right top edges of your drawing board with two 3mm machine screws each, either directly to the dry-erase board or through small pieces of project plywood. Mount the string pulleys to the motor shafts. I used thread bobbins from an old sewing machine; Amazon carries these as well. They're a slip fit on the 6mm motor shafts, so add a disc of masking tape first (Figure G). Wind about 4' of flexible braided fishing line from one bobbin to another. Try to keep roughly the same amount of line on each bobbin, and leave some slack between them. It doesn't matter which side of the bobbins the line exits.

Make the pen holder from ⅛" clear acrylic — scraps are often available at glass shops — following Figure 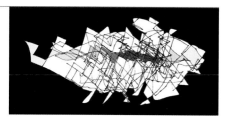, and hang it temporarily (upside down) on the slack "V" in your fishing line, for testing.

Connect the output of the RNG's Zener diode to any input pin of a microcontroller (Arduino, Basic Stamp, etc.). The L298N board has 4 inputs to control 2 motors, so I decided to grab groups of four consecutive signals from the RNG, such as 0010, 1010, etc., all using a single input pin of the microcontroller.

PROGRAMMING YOUR RANDOM DRAWBOT

You can download my code for the BASIC Stamp microcontroller from the project page at makezine.com/go/random-drawbot. But it's pretty simple to make your own version for your favorite microcontroller.

1. The L298N controls two motors, A and B. You must enable both by sending HIGHs to ENA and ENB on the L298N board (Figure F). Leave them enabled for the duration of the program. But just enabling the motors will not cause them to run; they must know which direction to turn.

2. Poll the RNG for a random combination of 4 zeroes (LOWs) and ones (HIGHs). I like to store them as variables (say, **W** through **Z**) to make it a bit easier to understand. If all 4 digits are (randomly) zero, go back and get 4 more. It takes at least one HIGH to move the pen.

3. The first 2 digits control motor A. **10** would turn the motor one way and **01** would turn it the other way. Similarly, the second 2 digits control motor B.

4. Send the 4 variables' values to input pins IN1 through IN4 on the L298N board.

5. Write subroutines to run the motors in their prescribed directions.

6. Add pauses or delays to keep the motor(s) running for whatever time it takes to draw distinct line segments. I use pauses anywhere from 100 to 500 milliseconds, but if you want to draw larger, um, creations, increase the delays to allow the motors to run longer.

7. Write a subroutine to brake the motors after they draw the line segments. LOWs to both IN1 and IN2 will brake motor A; LOWs to both IN3 and IN4 will brake motor B. You don't need to send LOWs to the motor-enable pins.

8. You can add another degree of randomness if you use an Arduino with the ability to control the motors' speeds with its PWM pins.

FIRST-TIME OPERATION

1. Connect a 4-wire cable from the L298N board to the motors. Polarity isn't important; it's random, you know.

2. Lean the easel back at 15° or 20° and do a dry run with just the pen holder hanging from the fishing line.

3. If the motors don't run or the pen holder's motions don't seem random, check your circuitry and program for errors. Shorten or lengthen pauses as necessary to compensate for motor variations.

4. Tape a piece of paper to the easel's drawing surface. Position the acrylic pen holder roughly in the center of the paper. You may have to manually wind/unwind fishing line from the bobbins.

5. Remove the top from your pen (I like to use a fine-tip Sharpie) and push it through the hole in its holder, so that the fishing line suspends the pen over the drawing surface (Figure ▓). Adjust the angle of the easel if necessary.

6. Run the program and stand back.

RANDOM TIPS

If the pen's moves don't seem random enough (the odds of the RNG spitting out, say, 4 zeroes 3 or 4 times in a row is very small!), and other attempts to fix them don't work, debug or print a series of the 4 random digits to your computer screen and pause awhile to study the numbers. Try changing the first transistor to another metal-can NPN transistor. It seems that

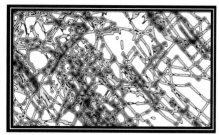

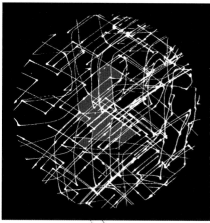

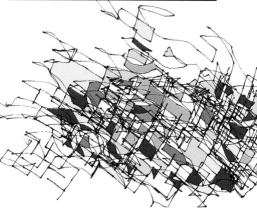

stressing the noise-producing transistor or running it on 12V instead of Platt and Logue's recommended 18V may not work with some transistors.

Random being random, the pen will sometimes try to go off the paper. It's OK to give it an occasional shove! I even added two switches and two more input subroutines to occasionally override any overzealous randomness.

Use your favorite photo editor to enhance your art: add or change colors, clone, copy, rotate, add, subtract, solid fills, textures. ◗

CHARLES PLATT is the author of *Make: Electronics*, an introductory guide for all ages, its sequel *Make: More Electronics*, and the 3-volume *Encyclopedia of Electronic Components*. His new book, *Make: Tools*, is available now. makershed.com/platt

**Time Required:
2 Hours
Cost:
$20**

MATERIALS

- » **Corrugated cardboard, 3"×6"**
- » **Scotch tape, ½" wide**
- » **File cards, 4"×6" (6)** or very stiff paper
- » **Wooden dowel, 12"×¾" diameter**
- » **Wood screw, 1¼" #8 size** Hex head preferred, flat-head Phillips will do.
- » **Aluminum tube, ¾" internal diameter, any wall thickness, 12" length**
- » **Neodymium magnet, cylindrical, ¹¹⁄₁₆" diameter, ½" long**
- » **Magnet wire, 30 gauge, at least 1 ounce (approx. 200 feet)**
- » **LED, red generic 5mm**
- » **Elmer's or epoxy glue**
- » **Fine sandpaper**
- » **Alligator patch cords (2)**
- » **Hookup wire, 22 gauge, 6" lengths, stripped at each end (2)**
- » **Aluminum channel, ½", 12" length (optional)**

TOOLS

- » **Utility knife**
- » **Ruler**
- » **Electric drill with ⅜" chuck**
- » **Drill bit, ⅛" diameter**
- » **Multimeter**
- » **Clamp or vise**
- » **Thumbtack or pushpin**
- » **Pen or pencil**

The Drifting
Magnet Mystery

Demonstrate the first law of thermodynamics with an aluminum tube and a magnet

Written and photographed by Charles Platt ■ Illustrated by Rob Nance

IN ONE HAND, YOU HOLD AN ALUMINUM TUBE 12" LONG. IN YOUR OTHER HAND, you have a small, polished metal cylinder. You drop the cylinder into the tube (Figure A), and — it disappears.

Where did it go? Nowhere! It's inside the tube, but instead of falling freely, it's moving slowly. After 5 seconds, it finally emerges at the bottom. It has just defied the force of gravity (partially, at least).

EQUIPMENT

To see this yourself, you'll need a cylindrical neodymium magnet measuring ½" high and ¹¹⁄₁₆" diameter. This is the smallest, cheapest option to generate a good result. I suggest you buy it from KJ Magnetics, which stocks the unusual ¹¹⁄₁₆" size. You also need 12" of round aluminum tubing with ¾" internal diameter (often abbreviated as ID). Many online sources such as Speedy Metals will sell you this for just a couple of dollars.

If you peek into the end of the tube after inserting the magnet, you'll see it mysteriously drifting down, as if it's falling through water. Aluminum is not magnetic, so why should this happen? When I demonstrated the phenomenon at the 2017 Maker Faire Bay Area, no one in the audience could figure it out.

EXPLANATION

To explain it, I suggest you make yourself a coil of wire that's just a fraction larger than the magnet. I described this experiment in my book *Make: Electronics*, but the version here is quicker and easier to build, and much cheaper, because I figured out a way to use a smaller magnet. The secret is to wind a coil that's a tighter fit.

Drill a ⅛" guide hole into the center of one end of the dowel, and drive the 1¼" screw into the hole until about ½" remains sticking out, as in Figure B.

Cut your file card to make a rectangle 6"×2⅜" as in Figure C. Fold the trimmed card in half (Figure D) and add tape to one edge (Figure E). Fold the tape over and open the card out as a tube (Figure F), and you should be able to slip it over the dowel (Figure G). If it won't fit, make another copy that is a fraction wider.

Now you need two circles of corrugated cardboard, shown in Figure H. To make each of them, first cut the outer edge, 2" in diameter, then draw around the end of the dowel in the center of the circle, and cut along that line. Punch a tiny hole in one circle, using a thumbtack or pushpin.

With the dowel inside the tube that you made from the file card, push the circles of cardboard over the tube and glue them ½" apart as in Figure I.

Now you need 30-gauge magnet wire. You can find it on eBay. Thread 4" of the wire through the little hole that you punched in one cardboard circle and tape it to the tube to stop it from waving around. Clamp a pen or pencil so that it is almost horizontal (sloping up a little) and place the spool of magnet wire on it. Wind a few turns around the card on the dowel, as in Figure J.

Tighten the chuck of your drill onto the head of the screw sticking out from the end of the dowel (Figure K), and you can use the drill to spin the dowel, taking up magnet wire from its spool.

You need about 600 turns of the 30-gauge magnet wire. If you put a piece of tape on the chuck of the drill, this will help you to count its rotations. Alternatively, if you have an accurate scale, weigh the spool of wire from time to time, and stop when you have transferred 1 ounce of wire from it onto the coil. See Figure L on the following page.

Don't wind the coil too tightly. You'll need to pull the dowel out, as in Figure M on the following page.

Use very fine sandpaper to remove insulation from the ends of the magnet wire. Attach an alligator patch cord to each end, and use a meter to check the resistance of the coil. It should be around 20Ω. If you can't measure it, or you get an intermittent value, you probably didn't remove the insulation entirely.

Allow your magnet to cling to the head of the screw in the end of the dowel, as in Figure N on the following page. Now for the big moment. Clip an LED between the patch cords, and push the dowel rapidly in and out of the tube so that the magnet energizes the coil (Figure O on the following page). You should see the LED flash.

EDDY CURRENTS

You've just demonstrated the way in which almost all electricity is generated in the world. The exceptions are electrochemical methods (batteries) and photovoltaic

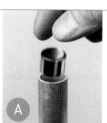

A 1¼" hex-head screw in the end of a ¾" dowel. You can use a regular screw, but the chuck of your drill won't grip it as securely.

The strip of card on the left is 2⅜" wide.

After cutting the card, fold it in half lengthwise.

Add Scotch tape along one edge.

After you fold the tape over, you can open the card to make a tube.

Slide the dowel into the tube.

Rings cut from corrugated cardboard.

Everyday Elmer's Glue will mount the rings on the tube.

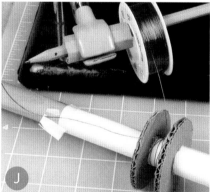
Preparing to wind magnet wire onto your tube.

The chuck of the drill grips the hex head of the screw.

A laboratory scale is a useful thing to have. You can find them affordably on eBay.

The completed coil consisting of about 1 ounce of 30-gauge magnet wire (600 turns).

The neodymium magnet attaches itself to the screw in the end of the dowel.

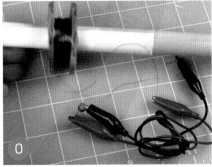
Pushing the magnet through the coil lights the LED.

A slot in an aluminum tube will reveal the magnet drifting slowly down.

methods (solar panels). If no one had ever figured out that a magnet can induce electricity in a coil, civilization as we know it would not exist.

Do you see how this relates to the Drifting Magnet Mystery? You can think of the aluminum tube as being like a very long coil with only one turn. Do you suppose that a moving magnet induces current in the aluminum? Tape a couple of pieces of hookup wire to the tube, spaced about 1" apart, and use a meter to measure millivolts. When you drop the magnet through the tube, or push it through, your meter should briefly measure a couple of millivolts.

In fact, the moving magnet induces an electrical flow known as *eddy currents* inside the aluminum. According to Lenz's Law, these currents create their own magnetic fields, which oppose the magnet.

But this still isn't the whole story. Remember that electricity creates a small amount of heat when it flows through a conductor. Where is this heat energy coming from? The magnet has potential energy, depending on its distance from the center of the Earth. As it loses potential energy by falling, the energy creates a tiny amount of heat in the aluminum. The first law of thermodynamics tells us that energy cannot be created or destroyed, and the Drifting Magnet Mystery demonstrates this.

DRIFTING TRICKERY

You can make the Drifting Magnet Mystery into a magic trick. All you need is a piece of metal rod that has not been magnetized but looks identical to the neodymium magnet. McMaster-Carr, or many other online metals vendors, should sell you a 1-foot length of $^{11}/_{16}$" rod.

Cut a section of rod equal in length to the magnet using a hacksaw or (easier) an abrasive disc in a handheld circular saw. Rub with fine sandpaper, and polish it. I'll call this the fake magnet.

Hold the real magnet in one hand, and keep the fake hidden in your other hand. Drop the real magnet through the tube a couple of times to demonstrate the drifting effect, then give the tube to a friend, quietly substituting the fake magnet. Needless to say, it will fall straight through.

Q The magnet will roll v-e-r-y slowly along a piece of ½" aluminum channel. Mysterious!

ENHANCEMENTS

I can think of a couple of enhancements. Figure P shows how I cut a slot in a piece of tube, using a hacksaw. This allows everyone to see the magnet as it descends.

Another option is buy a piece of ½" aluminum channel from a hardware store. You should find that the magnet fits into the channel with a tiny amount of clearance, so that it can roll without getting stuck, as in Figure Q. It won't move as slowly as it does through a tube, but eddy currents still prevent it from rolling as fast as you would expect.

If anyone asks you how this trick works, you can demonstrate the coil and the LED. Alternatively, just tell them that it demonstrates the first law of thermodynamics. That's what I told the audience at Maker Faire, although some people seemed a bit skeptical.

For more about magnetism and electricity, see my book *Make: Electronics*. And for a how-to guide on workshop tools, from saws to screwdrivers, you could look at my new book, *Make: Tools*. ◎

Shooting Killer Time-Lapses

Use these tips to wow your audience with beautiful, sped-up clips   Written by Mike Senese

To purchase, go to
www.joby.com

AS A HOBBYIST VIDEOGRAPHER, I have a fascination with how time-lapse videos reveal the graceful movement hidden in everyday life. When you start one, you never know what you'll end up with until you see the finished product.

Here are some tricks I've developed to help me get great results with my time-lapse videos.

» **Wide establishing shot.** You can't lose with shape-shifting clouds breezing over a bustling city or an open prairie. I try to use the widest lens I can, bordering on fisheye.

» **Unique perspectives.** Get above your terrain, or underneath your subject. My GorillaPod is great for letting me quickly set up high on a signpost or the ledge of a parking garage, offering vistas that aren't always seen.

» **Light changes.** The racing shadow play and color changes of a sunrise or sunset shouldn't be missed in a time-lapse. I often set my alarm for 45 minutes before sunrise to start my camera.

» **Strategic close-ups.** If you're going to edit clips together, it's nice to intersperse specific highlights. Use your storytelling skills!

» **High-quality optics.** Phones and action cameras often have time-lapse functions, but I try to shoot mine with a DSLR camera using the best lens I can access. Many interesting environments demand the image control this offers. Thankfully, the GorillaPod SLR-Zoom handles my heaviest gear with ease.

» **Good soundtrack.** A silent video is just fine, but the right song will make your video pop.

King Olav and the
Viking Sunstone

WILLIAM GURSTELLE's new book series *Remaking History*, based on this magazine column, is available in the Maker Shed, makershed.com.

Craft your own from polarizing film and locate the sun on overcast days

Written by William Gurstelle ■ Illustration by Peter Strain

Time Required: 2-3 Hours
Cost: $15

MATERIALS

» **Polarizing film, 3"×3" square piece** You can use polarizing sunglasses instead, but polarizing film is better and less expensive.

» **Cellophane tape, 4½"×½" piece** Cellophane is the film in which CDs are shrink-wrapped. This tape works the best, but Scotch brand clear packaging tape performed adequately for me. Be aware that other types of tape such as Scotch Magic tape do not work.

» **Basswood, 4"×4"×³⁄₃₂" piece**

» **Dimensional pine lumber, 2×2, 2" long** Actual size: 1½"×1½"×2".

» **Craft mirror, 3"×3" square**

TOOLS

» Saw
» Hole saw, 2¾"
» Electric drill
» Protractor
» Ruler
» Glue

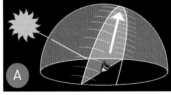

A

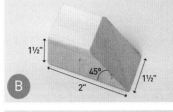

B

1½"
45°
2"
1½"

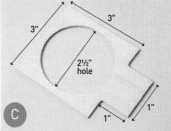

C

3"
3"
2½" hole
1"
1"

THE SAGAS ARE ANCIENT STORIES ABOUT THE SEA VOYAGES THAT THE VIKINGS TOOK AROUND THE YEAR 1000 CE. Most modern scholars believe the accounts are predominately fictional, but they contain enough factual details to give insight into the Viking way of life. One particularly interesting story concerns King Olav II, a famous warrior king who was so well thought of that he was canonized Saint Olav, the patron saint of Norway.

In the Saga of Saint Olav, Olav asks an Icelandic Viking named Sigurd to tell him the location of the sun during a foggy, cloudy day. Sigurd does so, and to check Sigurd's answer, the saga says that Olav "grabbed a sunstone, looked at the sky and saw from where the light came, from which he knew the position of the invisible sun."

That's quite a trick. And, if it were possible to have a device that could find the exact position of the sun on a cloudy day, it would explain a lot about how Viking voyagers were able to sail to far-off places such as Greenland, Iceland, and even North America. Historians and archeologists have found no evidence that the Vikings had magnetic compasses, and the North Star was of little use to Viking navigators because of the perpetual daylight of the arctic summer when most voyaging took place.

So, the Vikings had to use the sun to navigate on their long-distance journeys, but in the world's often-cloudy polar regions, days go by without the sun appearing. A *solarsteinn* (sunstone) that showed the sun's position on overcast days would be a great advantage.

There is no agreement among historians that sunstones actually existed, but there are tantalizing possibilities. First, sunstones are mentioned several times in different sagas. Secondly, the Vikings did know of naturally occurring rock crystals such as calcite and cordierite, and these minerals have the ability to polarize light passing through them under the right conditions.

HOW A SUNSTONE COULD HAVE WORKED

As any honeybee or mouse-eared bat will tell you, the sky is polarized. Regular light vibrates in more than one plane and is referred to as unpolarized. Polarized light vibrates in a single plane. With the possible exception of the Vikings, people did not know this until 1812 when the great French scientist Francois Arago invented the first polarizing filter. Thanks to Arago and the scientists that followed him, we now know that the sky's polarization is due to "scattering," meaning that light waves are bounced around by particles in the atmosphere.

The interesting thing about the sky's polarization is that it's not a random phenomenon but one that always occurs in the same way. If you could see polarized light like a honeybee does, you would notice a particular bright spot in the sky, and that spot is always in the same place: a point on the celestial arc that is exactly 90° opposite the sun (Figure A). If you can see that "spot" you can figure out where the sun is — even when it is occluded by clouds.

BUILDING A SUNSTONE

Only wealthy people like Olav could afford a sunstone made of rock crystal, but crafting your own sunstone with polarizing film is inexpensive and easy. The materials cost less than $15 and the project will take a couple of hours.

Juliann Brown, William Gurstelle, Hep Svadja

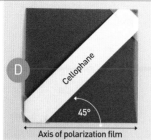
Cellophane
45°
Axis of polarization film

HOW TO MAKE A SUNSTONE

1. To begin, use a saw to cut a 45° angle at one end of the 2×2 piece of pine (Figure B).

2. Refer to Figure C to lay out and cut the 4"×4" piece of basswood.

3. Place the piece of cellophane at a 45° angle to the polarization axis of the film, which is probably the bottom of the square (Figure D). Hold it up to blue sky and rotate it. You should see the cellophane go from clear to opaque. If it does not, then: (1) you're not using the right kind of tape, or (2) you need to change the angle of the cello tape to the axis of the polarizing film, or (3) the cello tape has been placed on the wrong side of the polarizing film. Once correct, tape the cello tape to the film.

4. Align the polarizing film so that the cello tape faces out toward the sun. Rotate the film until the cello strip is perpendicular to the bottom edge of the basswood square (where the 1" tab protrudes). Then glue the 3"×3" polarizing film over the hole (Figure E). Trim film as necessary.

5. Glue the basswood piece with the polarizing film onto the square end of the 2×2 block, and the craft mirror onto the 45° face of the block as shown in Figure F. The film and the mirror are now at a 135° angle. Let the glue dry completely.

FIND THE HIDDEN SUN!

1. Hold the sunstone so that the cellophane strip runs vertically and the mirror is below the film holder. Make a guess as to where the sun is and then face the opposite direction. When you look at the sunstone you will note that the cellophane strip is darker (Figure G) or lighter (Figure H) than the surrounding area. Move the sunstone horizontally until the cellophane strip is nearly the same color as the polarized film on either side of it (Figure I).

2. Now, move the sunstone up or down until the cellophane strip completely disappears. You are now looking at a point exactly 90° on the celestial arc from where the sun is. Note: You need at least a small patch of blue sky to aim this sunstone.

3. Without moving your head or arms, look into the mirror. The spot in the center of the mirror is the location of the hidden sun. (Don't stare at the sun if it is visible when you do this! Cover the mirror with paper when testing your sunstone on sunny days.) You may need to adjust the angle of the mirror to get the most accurate readings. ◓

Time Required:
1-3 Hours
Cost:
$0-$3

MATERIALS

- » **Copy paper, cut to 4¼"×8½"** or similarly sized rectangle with sides in ratio 1:2
- » **LEDs (2)** Choose LEDs with medium to long wire leads. Test that they will work by sliding them both onto the battery.
- » **Coin battery, 3V CR2032 or similar**
- » **Aluminum foil tape, cut into 1"×¼" strips (2)** Found in hardware stores in the heating duct aisle. Copper foil tape (with or without conductive adhesive) can be used as well.
- » **Clear tape** Any kind of non-conductive tape can be used

KATHY CECERI's latest book is *Musical Inventions*. She is also the author of *Edible Inventions*, *Paper Inventions*, *Making Simple Robots*, and other books full of STEAM activities for kids and other beginners. When she's not busy writing, Kathy presents workshops for students and educators at schools, museums, libraries, and makerspaces throughout the Northeast. Visit her at craftsforlearning.com.

Written by Kathy Ceceri

Light Before You Leap

Press this origami frog's back to make its LED eyes glow and launch it into the air

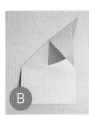

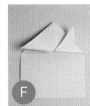

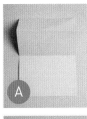

Hep Svadja

LIGHT-UP ORIGAMI DESIGNS HAVE BEEN AROUND FOR A WHILE — BUT THIS ADORABLE JUMPING FROG TAKES IT TO NEW LEVELS. Press the frog down, and the LED eyes start to glow. Release the frog, and it leaps across the table. If you're lucky, it may even do a little flip in the air!

The design for this light-up origami jumping frog comes from IBM engineer and STEAM education proponent Emi Olsson, who got the idea from the light-up paper art projects in my book *Paper Inventions*. When Emi sent me a video of her invention via Twitter after stopping by to say hello at a Mini Maker Faire, I knew I had to reverse-engineer it. It's since become one of my go-to projects for workshops and events, suitable for kids and adult beginners.

You don't need any origami experience — just patience, since there are some tricky parts that may take a few tries to get right. Once you've got your frog working, you'll need to (temporarily) dissect it so you can insert the LEDs. Building the circuit is a snap — all you need to connect the lights to the battery is a little bit of metallic foil tape. Then fold everything back up, and your frog will be jumping and glowing in no time!

1. PRE-FOLD THE PAPER

Bring the shorter edge at the top down to the bottom edge. Make the crease sharp.

Bring the top edge up to the middle fold and crease again. Open the paper up again (Figure A).

Bring one top corner down to the opposite end of the middle crease (Figure B). Crease and open again. Repeat with the other corner (Figure C).

Optional: Repeat Steps 2–4 with the bottom edge of the paper.

2. MAKE THE FROG'S HEAD AND FRONT LEGS

Take the X-fold on the top half of the paper and push in the sides (Figure D) to form a triangular "tent." Flatten the triangle (Figure E).

Fold the bottom corners of the triangle up, as shown in Figures F and G. Flatten.

3. MAKE THE FROG'S BODY

Fold up the bottom edge to the middle crease. Fold in the sides so they meet in the middle. You may need to lift the front legs out of the way (Figure H).

Fold the bottom up again so it meets the bottom corner of the head (Figure I).

4. MAKE THE BACK LEGS

Reach inside the last fold to grab the corner of one side (Figure J). Pull the corner out. Repeat with the other corner. The bottom now looks like a boat (Figure K).

Bring the corners of the "boat" down so they meet at the bottom, forming a diamond shape (Figure L).

Take one half of the diamond and fold it over so the edge meets the diagonal crease. Repeat with the other side to form the back legs (Figure M).

5. MAKE THE SPRINGY FOLD

Bring the bottom of the frog up along the middle crease, so the back feet are touching the front feet (Figure N).

Bring the same piece down so the bottom edge meets the middle crease (Figure O). Sharpen this fold.

Turn your frog over (Figure P). To give it a test jump, press down on the back edge to compress the springy fold. Slide your finger back to release it.

6. ADD THE LED EYES

Make sure the LEDs work together by sliding them both onto the battery. The positive wire leads (usually longer than the negative lead) must touch the positive (smooth) side of the battery.

Draw eyes near the frog's nose, then unfold the origami. Poke the LEDs through the eyes — making sure the positive (longer) lead is closer to the nose (Figure Q).

Take the foil strips and fold down one long edge so the glue sticks to itself. This will ensure that the metal foil makes a good connection between the LEDs and the battery. (You can skip this step if you're using copper foil tape with conductive glue.)

Inside the head, bend the bottom (negative) leads down so they're touching the paper and each other. Secure them to the paper with one piece of foil tape (Figure R).

Bend the top (positive) leads up and wrap the other piece of foil tape around them tightly. Place the battery on the frog, positive side up, over the foil tape. Use clear tape to hold it in place — making sure to leave the part of the positive side closest to the positive wires uncovered (Figure S).

7. TEST YOUR LIGHT-UP FROG

Bend the positive wires down until they're almost touching the positive side of the battery (Figure T). Fold the frog back up.

Time for the final test! When you press down the back, the LEDs should light up (Figure U). When you release it, the lights should go out as the frog leaps forward.

If the eyes stay lit, adjust the positive leads. Your light-up frog should last for many hops. ◐

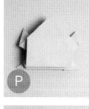
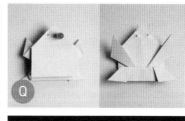

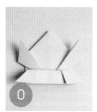

To see the jumping frog in action, and get a quick run-through of the origami steps, visit makezine.com/go/light-up-jumping-origami-frog.

Building Blocks

With the power of JavaScript and interaction with microcontrollers, coding tool **Make:Code** is fast, easy, and flexible

Written by Matt Stultz

MATT STULTZ is the 3D printing and digital fabrication lead for *Make:*. He is also the founder and organizer of 3DPPVD and Ocean State Maker Mill, where he spends his time tinkering in Rhode Island.

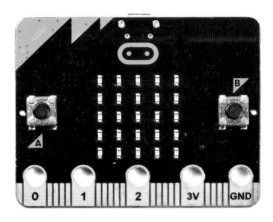

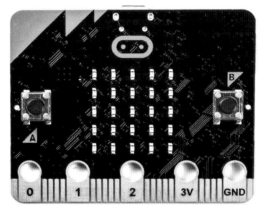

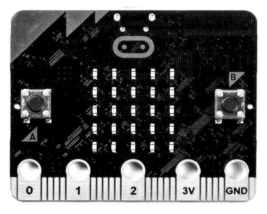

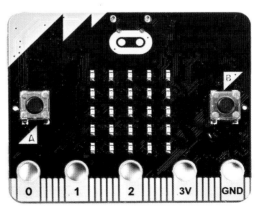

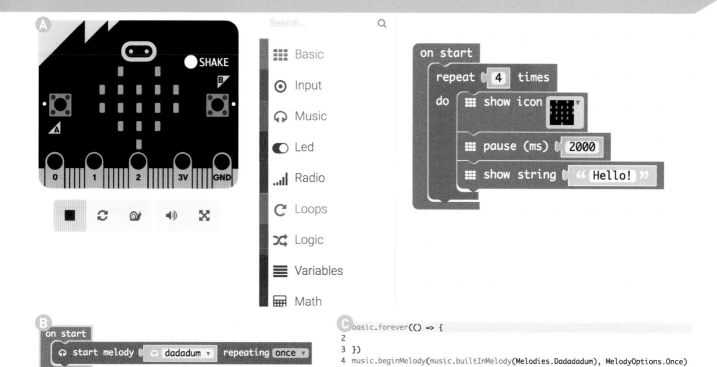

```
basic.forever(() => {
2
3 })
4 music.beginMelody(music.builtInMelody(Melodies.Dadadadum), MelodyOptions.Once)
5
```

WITHOUT A DOUBT, ONE OF THE MOST VALUABLE SKILLS IN OUR MODERN DAY WORLD IS BEING ABLE TO PROGRAM.
We are surrounded by technology; being able to program it bends it to your will. There are countless ways to get started being a programmer but few are as exciting as working with microcontrollers. These tiny computing devices allow you to not only interact with a virtual world, but also to reach out into the physical world.

Make:Code (makecode.com), a new coding environment from Microsoft, isn't the first block programming language, but it's the best implemented solution I have seen. These languages don't rely strictly on the user knowing extensive specific syntax but instead allow them to build applications by stacking commands together in a drag and drop interface(Figure Ⓐ).

Think about writing code from scratch like being a poet; you have the entire English language at your disposal, but finding the right words and making them flow is a skill that takes time and patience to master. On the other hand, block programming is like magnetic poetry stuck to your friend's refrigerator. Even the most mirthful partygoer can assemble a limerick or two from the available words to brighten their host's

cleanup efforts when found the next day.

One of the numerous features that helps Make:Code rise above the pack is that it doesn't just leave you stuck in block land — you can seamlessly switch back and forth to an actual code window. When you add a block to the program (Figure Ⓑ) and switch over to code view, the equivalent function displays in JavaScript (JS), the language behind Make:Code (Figure Ⓒ). Of course if you then make changes to that code in JS and switch back to the block window, the blocks will be updated with these changes. This ability to flip-flop helps those new to programming get started with blocks but quickly pick up JS syntax that can be used in writing applications outside of Make:Code.

JavaScript is a foundation technology of the modern internet. It allows developers to write rich applications that don't have to always rely on the server to do the computing but allow some of the work to be done locally in the browser. Gmail, Facebook, Makezine.com, even Make:Code's own site rely on JavaScript to be able to function. For those of you who are programmers and want to "nerd out," the Make:Code team even implemented a compiler that takes the code created by the user and prepares it for the chosen dev board in JS — no need for a trip back to the server.

BOARD SUPPORT
Designed for physical computing, there are a few boards currently supported by Make:Code, including the BBC micro:bit, Adafruit's Circuit Playground Express, the Chibi Chip, and SparkFun's SAMD21 dev board. While most of these systems are currently in beta, the micro:bit (microbit.org) is the best supported. This little powerhouse of a board is great for getting started. While your standard Arduino Uno has a single controllable LED built in, the micro:bit has a 5×5 LED matrix, built-in accelerometer (shake and movement), magnetometer (compass and metal detector), two buttons, light and temperature sensors, and even Bluetooth Low Energy radio. All of these together mean that a lot of projects can be completed without ever plugging in another component.

The implementation of Make:Code fully supports these components and, in the case of the radio, even adds some functionality to the board, allowing two micro:bits to communicate between each other without ever going through the standard pairing process required for most Bluetooth devices. The combination of Make:Code and micro:bit is easy to get started with and yet quite powerful; when I was selecting a platform to teach 200

Hep Svadja

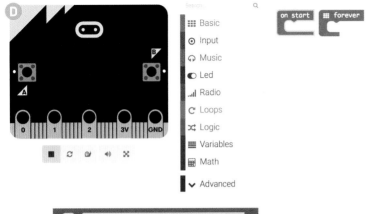

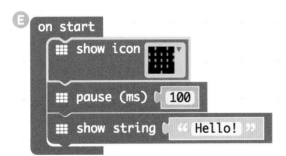

Mousing over the blocks gives you clear descriptions of their functions.

Create sequences of events by nesting code snippets.

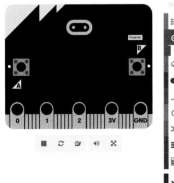

Each code function category offers multiple options to let you make intricate code creations.

Saudi Arabian teenagers about working with microcontrollers, I chose this pair.

Getting started with Make:Code and its supported platforms is easy. There are no downloads, no drivers, no real setup. Make:Code runs entirely in browser, so you only need an internet connection to use it. Thanks to the built-in simulator, you don't even need an actual board.

MAKING CODE

Let's test it out. First, go to makecode.com and select the board you want to use. I will be using the micro:bit for all examples here. Once the new project is open, a simple interface will greet you with a simulated dev board on the left, a list of commands in the middle, and a code window containing your first two blocks on the right (Figure D). These starter blocks will be vaguely familiar to anyone who has ever done any Arduino programming before; "Start" and "Forever" are the Make:Code equivalent of "Start" and "Loop" — these two functions are the base of every application. Any code that is added to the "Start" block will execute whenever the device is turned on, restarted, or reset. This makes it easy to add code that you only want to execute once for things like setting the initial position of a motor or turning on a sensor. The "Forever" block executes after start but runs over and over without stopping until the device is turned off or reset (at which point start will run, followed by forever again).

Adding code to these first two blocks is as simple as dragging and dropping. Between the code window and the simulation, you will find the commands to build your application. These are broken up into groups of like commands. For the micro:bit this includes Basic, Input, Music, LED, Radio, Loops, Logic, Variables, Math, and Advanced (I will leave you to explore that on your own). If you don't find the command you are looking for in a given group, don't forget to click on the "more" button. Also remember to scroll through the options as some might be hidden off your screen.

On the micro:bit, Make:Code allows you to access the LED array in a variety of ways that take much of the work out of using it. Not only can you toggle each LED

micro:bit, Hep Svadja

individually, but you can also show or scroll full words, sentences, and icons across the matrix. All without the user needing to create a buffer or individually controlling each LED in the matrix (Figure ⓔ).

STACKING BLOCKS

The Make:Code blocks snap together, stacking to make the final application. Some blocks fit inside others and have shapes that help indicate this. One complaint I have is that sometimes the block shapes are too similar; it's hard to know which block will or won't work. I would love a system where clicking on a block or an empty socket will change the available items to only show those options that can be used with the item that was clicked. This would be very similar in my opinion to how intellisense works in Microsoft's professional development tools.

Each time a block is added or a variable is changed, the simulation window automatically updates, showing how the code will work on the board. I was really impressed the first time I was shown that not only will it do so with the components

on the board, but it will also show you how to hook up basic external components to help the user also learn circuits. I envision a future where more sensors and outputs are supported, giving the user a great way to develop their entire project virtually before ever buying a single piece of hardware.

Make:Code works great on mobile devices too!

Once complete, the compiled application can be downloaded to your computer or uploaded directly to the device. Make:Code-compatible devices show up as flash storage drives on your computer, meaning they do not require drivers to get working. All that is needed to program the device itself is to copy the *.hex* file that downloaded from the page to the device while it is plugged into USB. Drag, drop, done. The device will then begin running the code that was copied to it.

If I had to leverage any real criticism against the system, it would be that it makes the device a bit unresponsive. This

is probably something someone new to using microcontrollers would never notice but to those familiar with their real-time snappiness, the extra weight of all of the code that makes things easy, makes the end device slow.

OUTLOOK

I'm really hoping that Microsoft commits to Make:Code and continues to improve it. For those of you who are looking at getting started with microcontrollers and especially those of you who are educators and might be working with large groups, give the blocks a shot, you might find they help you build tomorrow's software architects. ⊘

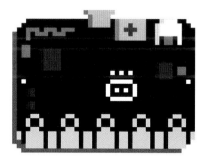

Helpful Hints for

Written by Aaron Newcomb

Linux

Save time and eliminate errors with these two command line tricks

AARON NEWCOMB has been using Linux since 1997. He has worked in the IT industry for companies like Oracle and Hewlett Packard. He co-hosts shows about technology for TWiT, including FLOSS Weekly, This Week in Google, and The New Screen Savers. In 2012, he founded Benicia Makerspace where he serves as president and executive director.

This is an excerpt from Aaron Newcomb's book *Linux for Makers*, available on makershed.com and fine book retailers everywhere.

LINUX IS A POWERFUL OPEN SOURCE OPERATING SYSTEM THAT HAS BEEN AROUND FOR MANY YEARS and is widely used for running servers and websites. But most students and makers encounter it for the first time when they are working on projects with their Raspberry Pi or similar single-board computers (SBCs) such as BeagleBone Black or Intel Galileo. By gaining a deeper understanding of Linux, makers can add another useful tool to their kit that will help them build their projects more easily.

If you are like me, your spelling and typing abilities may be lacking. Too many times I have spent 20 or 30 seconds typing a long command with lots of options only to find out after I hit enter that I had something wrong and needed to start from the beginning again. Not only that, but with all the possible choices, it can be hard to remember exactly the command you used to perform a certain task from day to day. Luckily, the Linux shell has some tools built in that can help with both of these problems.

AUTO-COMPLETE A COMMAND: TAB

You can use the auto-complete feature of the shell by simply pressing the Tab key on the keyboard. This will auto-complete a command that has been partially typed and it will also auto-complete a filename based on the context of what you are typing.

For example, if you type "**tou**" and press the Tab key, the shell will fill in the rest of the missing letters to make "**touch**". If there are multiple options that start with the letters you have entered, the first time you press Tab nothing will happen. If you press it again, however, the shell will display a list of all possible commands or file names that start with the letters you entered. So, if you type "**mkd**" and press Tab twice, you will be presented with two options for commands that start with mkd: **mkdir** and **mkdosfs**:

```
pi@raspberrypi ~ $ mkd
mkdir mkdosfs
pi@raspberrypi ~ $ mkd
```

If you continue to add more characters and then press Tab, you will eventually rule out all the other options and the shell will complete the rest of the command or filename when there is only one choice left.

This auto-complete feature is a real time saver with bigger commands and long file names. It also eliminates spelling errors when you haven't used a command very often yet.

SEARCH FOR A PREVIOUS COMMAND: UP, CTRL-R

Linux keeps a history of all the things you type into the command line. A simple way to review the commands you have typed

TIP: *By default, Tab doesn't always know about the available options for a command, but can auto-complete the name of the command and any associated file names that might be used as part of a command.*

is simply to use the Up Arrow to scroll back through each command starting with the most recent. If the command you are looking for is further back in your history, you can search for it by pressing "Ctrl-R" on the command line followed by some characters. For example, if you wanted to search for the last time you used **nano** to edit a file you could press "Ctrl-R" followed by "**nano**".

It doesn't matter if there's already some information entered at the cursor when you press Ctrl-R. That text won't be used for the search, only what you type after you press Ctrl-R. Notice that the prompt changes to (**reverse-i-search**) followed by the letters you entered when doing this type of search through your command history.

```
(reverse-i-search)'nano': nano
hello.sh
```

If you press one of the arrow keys, Home, End, or Tab, you will finish the search and be able to edit the command that you looked up. You can also continue to search through your history by pressing Ctrl-R multiple times before you exit out of the search.

TRY IT FOR YOURSELF:
Change to your home directory and create a file by typing:

```
cd
tou <TAB> file1
```

When you press Tab it should complete the name of the **touch** command. Now change to your Downloads directory by typing:

```
cd D <TAB> <TAB>
```

You should see something similar to this:

```
pi@raspberrypi ~ $ cd D
Desktop/     Documents/
Downloads/
pi@raspberrypi ~ $ cd D
```

Add the letters "**ow**" and press Tab again to auto-complete the path we want and press enter.

Now let's create our second file by using the command history. Press Ctrl-R followed by "**tou**":

```
pi@raspberrypi ~ $ cd D
Desktop/     Documents/
Downloads/
pi@raspberrypi ~ $ cd
Downloads/
(reverse-i-search)'tou': touch
file1
```

Press the End key and change "**file1**" to "**file2**". Press enter to complete the task. Now you have created two files — one in your home directory and one in the Downloads directory. You have also saved a lot of typing in the process! ◐

Hep Svadja

TOOLBOX GADGETS AND GEAR FOR MAKERS

L-CHEAPO LASER ENGRAVER
$195–$595 endurancerobots.com

When you're a proud owner of an open source 3D printer, you have a lot of options on how to configure and upgrade it. A popular option is to upgrade your tool head, but why stick with an extruder when you can add a laser? The L-Cheapo Laser Engraver kits, offered in 2.1W–8W options, can be added to any open 3D printer or CNC machine. This is a diode laser system; these have gained popularity since Blu-ray players brought the price down on high-power, solid-state lasers. While not as powerful as a real laser cutter, the L-Cheapo is good at lightweight tasks like etching and cutting thin materials. I tried the L-Cheapo on a 3D printer, but I'm looking forward to slapping it on a big CNC to have fun with large scale etches.

Be careful with this one. It comes with safety glasses, but that is the only safety system in place. The laser is more than powerful enough to blind you. Aimed in the wrong direction, it could easily start a fire, too. Don't use the system without wearing the safety glasses, and be sure to have a fire extinguisher nearby.

The name of this machine may have more to do with the construction quality than the price itself. At nearly $200 for the 2.1W, I would love to see a higher build quality than what the L-Cheapo provides. Still, if you want to add a laser to your G-code-running tool of choice, the L-Cheapo is a great option. —*Matt Stultz*

Endurance

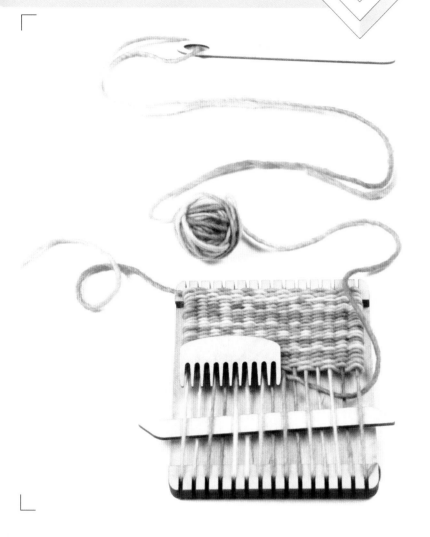

PURL & LOOP LOOMS
$20–$110 purlandloop.com

Simple, portable, and just plain fun, the looms from Purl & Loop are perfect for busting through your yarn stash. And the best part? They're easy to use regardless of skill level, from beginners to seasoned weaving pros.

These little looms and loom kits range in price from $20 to $110. Size and purpose run the gamut from tiny swatch testers, to bracelet looms, to place mats. Maker-made in Texas, the kits include tools you need for completing your project as well — some include pickup sticks, a carrying bag, or a tapestry needle for finishing, plus concise yet very thorough step-by-step instructions.

I had the pleasure of testing three of the kits: the Minute Weaver Set, the Wee Weaver Kit with Micro Snips, and the Stash Blaster Bracelet Loom Starter Package.

All were fun to work with, and set up was quick and easy. As an added bonus, the smaller looms can function as jumping off points for much larger projects. Why not turn a set of swatches into a table runner, tote bag, or throw? —*Mandy L. Stultz*

MAKER MUSCLE
$99 for Basic Kit Kickstarter Reward makermuscle.com

Making things move in a circular motion is easy. Turning that circular motion into linear motion can get tricky. To help makers build their dream projects, Diego Porqueras of Deezmaker created Maker Muscle, an easy-to-use linear actuator. The power behind Maker Muscle comes from stepper motors like those commonly found in 3D printers. This means you can control the actuator with electronics that have become affordable and easy to find.

The case of each Maker Muscle is made from aluminum extrusion and provides plenty of mounting options with attachment channels running down all four sides. I was inspired by the potential projects the unit could fit into — lots of Halloween props, robots, and home automation builds are going to be sporting Maker Muscles in the near future! —*MS*

MACCHINA M2

$89 macchina.cc

Wanna hack your ride? The Macchina M2 is essentially a programmable, Arduino-compatible microcontroller board with an OBD2 interface to communicate with your car's CAN bus. Want to get full access to everything your car is communicating? Easy peasy. Want to engineer that into your own custom dashboard or data logging? Doable! Want to define your own fuel mapping? Should be possible. Full autonomous? Maybe ... and if you do, let us know!

In our testing we went as far as expressing the classic Arduino blink sketch by activating our hazard flashers — we considered mapping that same output to the throttle control, but thought better of it. Hacking using Macchina isn't as straightforward as other microcontrollers; the CAN bus has been standardized as an interface, but not as a language. Fret not, as there are plenty of open source tools available to you, like GVRET (the General Vehicle Reverse Engineering Tool) and SavvyCAN, a software tool that helps isolate and interpret the codes your car is sending. Once you've found the codes you need, you can send them back just as easily — and there are tons of tutorials on the Macchina website to get you going.

—Tyler Winegarner

EIGHTBYEIGHT BLINKY

$45 blinkinlabs.com

We've all seen grids of RGB LEDs, but the EightByEight sets itself apart with its features and style. Spotted at a table in the Dark Room at Maker Faire Bay Area, I knew I had to take *Make:* editor Caleb Kraft over to see it. The EightByEight is an 8×8 grid with a built-in controller, Wi-Fi, a rechargeable battery, and an accelerometer.

For Caleb, the tipping point was the simple interface for programming patterns. The Blinkinlabs team has created a drag and drop configuration utility that allows you to draw patterns and animations — you can even drop in a bitmap and send it to the board. Four mounting holes and an included lanyard make it easy to wear it around your neck. Walking around Maker Faire with an EightByEight mixed in with your badge is sure to get you some attention. *—MS*

KANO PIXEL KIT

$80 kano.me

If you have an inquisitive young child in your life who wants to get into electronics, the Kano Pixel Kit is a great tool to introduce the fundamentals of software and hardware.

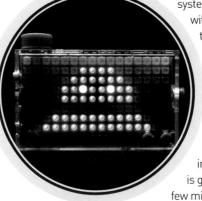

The intuitive, puzzle-block-style code system teaches programming, with JavaScript available under the hood once you master the fundamentals. Tutorials via a gamified challenge tree teach how to use sensors such as accelerometers and microphones to trigger events, as well as creating simple custom animations in the Kano app. The Pixel Kit is great for adults too. In just a few minutes I was able to design a simple marquee scrolling script with a sound trigger to remind the office to be quiet during our *Make:* live streams. Kids can share scripts and animations with friends, and a variety of other add-ons such as a camera kit and a motion sensor kit are coming soon. *—Hep Svadja*

BUILDTAK FLEXPLATE SYSTEM

$40–$190 buildtak.com

BuildTak is one of the top choices for build surfaces among 3D printing enthusiasts. If there's one complaint you hear about it, it's that it can work too well and it's sometimes hard to get your prints off the bed. To simplify this process, BuildTak has introduced their FlexPlate system. The FlexPlates are spring steel plates held in place with magnets that are embedded in a secondary plate that adheres to your printer. BuildTak is added to this spring steel plate and after the print is complete, the user can remove the plate and lightly bend it to force the print to pop right off.

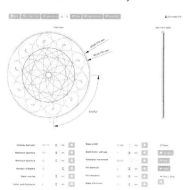

I have used the system on a couple printers and love it. I suggest picking up an extra steel plate so you can swap between them, which brings the downtime between prints to a few seconds. BuildTak was not the first to come out with a system like this, but their implementation is one of the best we've seen. If you are a 3D printing pro, removable plates are a step toward automation, and the FlexPlate system can easily be added to your current farm. —*MS*

IRIS CALCULATOR

$5 monthly, or $35 yearly iris-calculator.com

Last year, I found myself obsessing over making a mechanical iris peephole for my front door. I worked up a prototype over the course of a month or two and learned a lot. During the build process, I discovered The Iris Calculator by Matt Arnold — it's a fantastic tool and exactly what I needed to tweak my design.

Designing irises from scratch isn't insanely complex, but it isn't particularly fun either. Iris Calculator took the pain away. I could switch major design choices like number of blades or maximum width in an instant and have a new design. This saved me considerable time.

I entered the measurements I wanted and it spit out a DXF file for me. I was able to take this file into CAD and extrapolate the design from there. Ultimately, the cost of the software was almost nothing compared to the time I would have spent re-designing to try different geometries. I'd highly recommend it to anyone wanting to make their own mechanical iris. The only caveat is that it only has one style of iris right now, but Arnold plans to develop the app with new features.

—*Caleb Kraft*

PRESERVATION:
THE ART AND SCIENCE OF CANNING, FERMENTATION AND DEHYDRATION
by Christina Ward
$25
processmediainc.com

Christina Ward is more than a Master Food Preserver — she's a born teacher with a punk-rock, Midwestern socialist, DIY attitude. I warned her I had no time to read her book then proceeded to read it anyway because it's so much fun — a witty tour de force of food-nerd physics, chemistry, and history, with snappy how-tos for preserving with sugar, salt, acids, heat, pressure, microbes, dehydration, and smoke. Ward shares how and why these techniques work, with hundreds of pages of recipes: fruit jams and pie fillings; a panoply of pickles, chutneys, and relishes; sauces from pomodoro to habanero to chimichurri; fermented kraut, kimchi, kombucha, and kvass; even low-acid foods like meats, veggies, beans, and soups. This book deserves a place in every DIY pantry. You'll refer to it often and be a smarter, safer, and more capable cook.

—*Keith Hammond*

THE TOTAL INVENTOR'S MANUAL
(POPULAR SCIENCE): TRANSFORM YOUR IDEA INTO A TOP-SELLING PRODUCT

by Sean Michael Ragan
$29 weldonowen.com

The Total Inventor's Manual offers something for anyone who's ever considered making or selling their own product. Do you have ideas? Or need them? Are you trying to make a product? Need advice about how to spread the word? It points you in the right direction, and shows you what you need to do.

I hate to call this a reference book, because it doesn't read like one. The visuals are fantastic, and the writing style keeps things interesting. Historical anecdotes and inspirational stories complement the "lessons" nicely.

If you have yet to make or sell your product, this book will teach you useful things. Even if you have, you'll learn a thing or two. —*Stuart Deutsch*

E3 CNC ROUTER KIT

A low-cost, build-it-yourself machine that doesn't sacrifice features
Written by Chris Yohe

THIS WOOD-CUT CNC KIT FROM BOB'S CNC LOWERS THE BARRIER TO ENTRY significantly, while upping the ante on included elements. For less than $600, you get a reasonably sized desktop router, plus everything you need to start, including both collets (⅛" and ¼"), a threaded work surface, and work clamps.

BITS AND PIECES

Bob's CNC lays out clear step-by-step instructions for construction. Builders should plan on tackling this as a weekend project. Beginners may want to do a bit of research (tape is your friend). Most of the parts are mirror images, so you rarely need to worry about orienting a part the wrong way.

The kit includes enough cable wrap to clean up the wiring, plus instructions on how to route the wiring to avoid normal pitfalls like accidentally triggering the endstops (which are also included) with interference. You'll need to tether your machine to a computer, but since the software is OS agnostic, you can use something as simple as a Raspberry Pi.

The recommended software is all fairly common and open source. F-Engrave and its associated tutorials will have you up and cutting shortly, and a few custom tutorials will help in setting up software to match the machine. Universal G-Code Sender is quick to set up, but has a learning curve. After a few homing

and zeroing miscues, your works will be coming off the machine almost as fast as new ones are sent to it.

HELP WANTED

Tutorials are posted regularly to the Bob's CNC website, but more are needed to help beginners get up to speed safely and with slightly less frustration. While the recommended software is great for sign and logo engraving (and even inlays), it's lacking in certain aspects. Other software such as Easel can be used to generate the G-code, if the user prefers that.

After tweaking our feeds and speeds we got good results, but we had to be conservative. While the wood-cut construction was more rigid than we expected, it did suffer from some flex after we pushed it too far, as well as deformities in the work, so we stuck with softer materials. It's a trade-off — the flex actually saved a bit or two of ours.

THE PRICE IS RIGHT

For those who want a simple, affordable CNC router, this machine is great. We look forward to seeing more documentation and to pushing this machine further. At this price, it's time to get your hands a little dirty, and have a lot of fun! ⊘

- **BASE PRICE** $588
- **PRICE AS TESTED** $588
- **ACCESSORIES INCLUDED AT BASE PRICE** ⅛" end mill, threaded work surface, DeWalt DW660 router, ¼" and ⅛" collets, set of workpiece clamps
- **ADDITIONAL ACCESSORIES PROVIDED FOR TESTING** None
- **BUILD VOLUME** 450×380×85mm (17.7"×15.3"×3.3")
- **MATERIALS HANDLED** Wood
- **WORK UNTETHERED?** No
- **ONBOARD CONTROLS?** Power switch on router, no emergency stop
- **DESIGN SOFTWARE** F-Engrave
- **CUTTING SOFTWARE** UGS (Universal G-Code Sender)
- **OS** Windows, Mac, Linux
- **FIRMWARE** Grbl
- **OPEN SOFTWARE?** Yes (F-Engrave is FOSS, UGS is GPLv3)
- **OPEN HARDWARE?** No (Custom design hardware. Uses open source CNC Shield, Arduino)

bobscnc.com

PRO TIPS

1. Amazon or the usual outlets are your friends for end mills. To start, nab yourself a 60° V-carve bit and a flat cut bit.

2. Don't forget: *Make: Getting Started with CNC* will be a great help. makershed. com/products/ make-getting-started-with-cnc-1

3. Take it slow to start and get yourself some hearing protection!

WHY TO BUY

If you've been wanting to get into CNC routing, but were waiting for the right price, Bob's CNC has you covered.

CHRIS YOHE is a software developer by day, and a hardware hacker by night. The Chief HotEnd of 3DPPGH and a member of HackPittsburgh, he is an avid 3D printing enthusiast. From rugby, to tailgating, to 3D printing he's always looking for an excuse to make the world a better, or at least more interesting, place.

Hep Svadja

ome Suite ome

Personalize your space with **Make:** books

DESIGN FOR CNC
**By Anne Filson and
Gary Rohrbacher $35**
Why buy one-size-fits-all
particleboard furniture
from halfway around the
world when a CNC machine
empowers us to locally
fabricate designs made
anywhere, using whatever
local materials we choose?

GETTING STARTED WITH 3D CARVING
By Zach Kaplan $20
Faster than 3D printing,
with a wider choice of
materials, 3D carving
creates durable,
permanent parts that look
great. Learn the basics
of designing and making
things with a 3D carver
and build a guitar, clock,
earrings, and even a
skateboard.

GETTING STARTED WITH CNC
**By Edward Ford
$25**
If you've wanted to add
a CNC router to your
workshop but weren't
sure where to begin,
this is the book for you.
No prior knowledge or
experience is required
— you don't even
need CNC access to
complete the exercises!

AVAILABLE AT MAKERSHED.COM AND FINE RETAILERS EVERYWHERE

What's New:

Linux for Makers
By Aaron Newcomb
$25
Open source making.

Musical Inventions
By Kathy Ceceri
$25
Help kids learn that music is everywhere.

Making Things Smart
By Gordon Williams
$35
Use Espruino to make everyday objects into intelligent machines.

Making Things Talk, 3rd ed.
By Tom Igoe
$35
Third edition of the seminal sensors book!

Modern Leatherwork for Makers
By Tim Deagan
$30
Steampunk meets digital fabrication — and beyond.

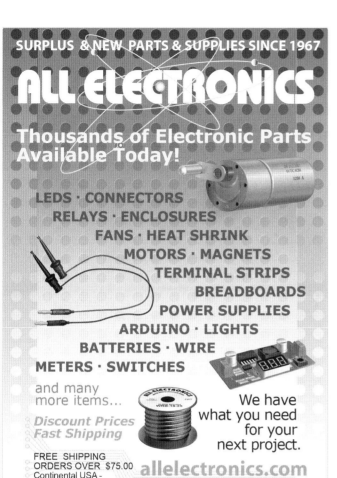

SHOW&TELL

Dazzling projects from inventive makers like you

Sharing what you've made is half the joy of making! To be featured here, show us your photos by tagging #makemagazine.

1. **Michael Blunt** built this teardrop trailer for his dog. He also sells them on Etsy and shares instructions to build your own at makezine.com/go/puppy-camper.

2. Instructables user **Stephen Taylor** (@buck2217) constructed this adorable igloo from 2×4s for his goats to frolick on.

3. The fish in Imgur user **beatdownllama**'s aquarium look right at home swimming with the snowspeeders on Hoth's icy tundra.

4. **Jillian Northrup**'s cat transit system is purrfectly pipe fitted up and around the angled ceiling.

5. Rocky the dog is quite pleased with this tropical piece of real estate built by Imgur user **iwanebe**.

6. Kitty is sure to live long and prosper with a Starship Enterprise themed perch, thanks to **Twisted Tree Pet Furniture** (@twistedtreepet).

7. **Matt Macgillivray**'s furry best friend can enjoy the harbor in style with this doggie houseboat.

8. Through their CatastrophiCreations Etsy shop, owners **Megan** and **Mike** offer you the chance to transport your cats back to the 1950s with this vintage TV cat bed.

9. **Ben Uyeda** of Homemade Modern provides instructions to make this sleek and simple geometric doghouse at makezine.com/go/geometric-doghouse.